Networking for Nerds

Networking for Nerds

Find, Access and Land Hidden Game-Changing Career Opportunities Everywhere

Alaina G. Levine

WILEY Blackwell

Published by John Wiley & Sons, Inc., Hoboken, New Jersey
Published simultaneously in Canada

For general information on our other products and services or for technical support, please contact our Customer Care Department within the United States at (800) 762-2974, outside the United States at (317) 572-3993 or fax (317) 572-4002.

Wiley also publishes its books in a variety of electronic formats. Some content that appears in print may not be available in electronic formats. For more information about Wiley products, visit our web site at www.wiley.com.

Library of Congress Cataloging-in-Publication Data:

Levine, Alaina G.
  Networking for nerds : find, access and land hidden game-changing career opportunities everywhere / by Alaina G. Levine.
      pages cm
  Includes bibliographical references and index.
  ISBN 978-1-118-66358-5 (pbk.)
  1. Business networks. 2.   Career development.   I. Title.
  HD69.S8L475 2015
  650.1'3–dc23

                                                            2014049388

Cover image: Photo by Pete Brown
Printed and bound in the United States by RR Donnelley.

2   2015

Dedication

This book is dedicated to my mother, Susan Levine, who taught me to always look for, seek out and ask for opportunities, no matter the perceived obstacle, and whose love and confidence in my abilities has helped me achieve my wildest dreams; and to my brother Joshua Levine, whose support, love, and laughter has ensured I remain sane along our road together.

"Alaina G. Levine is a Networking Ninja. I've learned a lot from her that has helped me get where I am today, and if you follow even half the advice in this book, you'll be networking better than most scientists I've met. You'll immediately see direct, tangible benefits for your career."

Dr. Kevin B. Marvel
Executive Officer
American Astronomical Society

If you want a great job, if you want to forge new professional connections, you'll need to network. Anxious about getting started? Don't be. Once you've read Alaina G. Levine's new guide, you'll discover that networking is natural, effective – and even fun!

Charles Day
Online Editor
Physics Today

This wonderful book is for both those who are new to networking and those who are seasoned networkers. You will learn novel techniques to help you navigate and succeed in the professional world and open your eyes to new career directions. It provides a down-to-earth, common sense approach to networking and will ultimately help you achieve your career goals. I highly recommend *Networking for Nerds!*

Michelle Horton, CMP
Director of Administration and Meetings
Ecological Society of America

Alaina G. Levine has provided thousands of AGU's student and early career members with invaluable advice during her webinars, workshops, and one-on-one consultations. With the release of *Networking for Nerds*, scientists around the world will have access to Alaina's real-world experience and expertise in the comfort of their own homes and offices.

Chris McEntee
Executive Director/CEO
American Geophysical Union

Contents

Foreword

Networking is the Essence of Research

To many budding researchers the thought of networking brings about visions of unsavory representations in smoke-filled rooms – of prostituting one's scientific ideals to get ahead in the world. My experience is that networking is instead the essence of scientific progress, and should be embraced as one of the reasons why we choose to do research as a career.

The author and Nobel Laureate Brian Schmidt, Copenhagen, 2014.

Humanity is a small part of a small planet, which is only 1 of 100 billion planets in our own galaxy, the Milky Way, which itself is only 1 of 100s of billions of galaxies in the visible universe. Yet, through the scientific process, over the past 400 years we have managed to build a comprehensive understanding of the cosmos from the sub-atomic particles that make up normal matter, to the universe on its largest scales. We are insignificant, yet our knowledge is able to explain the vast scales of the universe from the first few moments after the Big Bang to its current state 13.8 billion years later. This knowledge has been gained by building on the toils of previous generations of scientists, working sometimes competitively, but always collectively, towards furthering knowledge. Networking is all about the connections that enable science to progress.

While many of us might like to work on problems in isolation, ask yourself: "If I make a discovery and shout it out into an empty forest, and no one hears it, have I made a discovery at all?" You of course have, but humanity has not – it is only by sharing what you have learned that science benefits.

Networking is not just about sharing what you have learned; it is also about contemplating what is possible. In 1994, while visiting Chile, I found myself in the office of Nick Suntzeff, a scientist who served as one of my scientific mentors during my thesis. We were discussing the prospects of measuring the ultimate fate of the universe, by using the work he and his colleagues had recently completed on supernovae and newly released technology in the form of large digital cameras. It was on that day we hatched the plan that turned into the High-Z SN Search, the discovery of an accelerating universe and, 17 years later, a Nobel Prize in Physics.

Networking is an important part of the scientific process and, therefore, doing it well is an important part of being a successful scientist. The innate skills each of us has in networking vary widely, but as with other skills, most of us can improve with training. This book is all about the basic skills you need to learn to better communicate with your colleagues. While much of what you learn has the indirect benefit of improving your career prospects, the primary benefit of learning to network successfully is that it will make you a better scientist. And that is something everyone can be proud of.

Brian Schmidt
2011 Nobel Laureate in Physics

Introduction

The book you are reading right now is a direct result of networking. In 2011, I was writing an article for *Science Magazine* about common mistakes to avoid in the postdoc appointment and was looking for sources to interview. I asked a colleague if she knew of anyone and she introduced me to Sarah Andrus, an editor at Wiley who was overseeing a program called Wiley Science Advisors. Sarah transmitted my request for potential interviewees to her international network of Advisors and within mere hours I had more sources that I even needed. I immediately recognized that there was synergy between our interests and projects (as did she), and we began a series of discussions in which we learned about each other's goals and expertise and examined avenues for collaboration. This led to my giving a networking webinar as part of the Wiley Learning Institute in 2012, and another discussion with Sarah in August of that year which led to the birth of this book idea.

If I analyze my career as a series of opportunities – opportunities that I have either found, been told of, asked for, been offered, or created myself – I can easily see a direct path that has led me to where I am today. And every opportunity was a direct result of networking. Networking has literally gotten me jobs, awards, opportunities to sit on boards and committees, speaking engagements, invitations to review grant applications, invitations to apply for grants and fellowships, and now this book. Networking has directly provided me with new knowledge about my various professions and suggestions for taking my career in novel directions. Networking has given me access to people, places, information, and inspiration that have transformed my career and my life in ways I couldn't have even imagined.

That's how powerful networking is.

In fact, if there is one message that I want to emphasize, amplify, and continuously shout from the rooftops via this book it is this: Networking is *the key* to career advancement. Even today, after 13 years of what I refer to as "hard core networking," in which I have dedicated a focus to building and cultivating mutually-beneficial networks, I am still amazed at the riches networking has been able to bring me.

Networking can continuously pay dividends, but most people, especially scientists and engineers, don't realize that to reap the benefit of networking, they have to invest time and energy into it. Unlike business majors, who are taught to network from day one, nerds like myself, who studied science and engineering in college, don't get schooled in the ins and outs of networking, and that it is an absolute necessity to progress in their careers. If we are lucky, we learn about networking by watching others in our profession do it, such as our advisors. But more often than not, scientists and engineers perceive that networking is not a core function of moving forward, and in fact it is often erroneously seen as a sideline activity that takes time away from their preconceived notion of the only avenue to get a job – to do good science or engineering, which is demonstrated in outputs like publishing, presenting at conferences, and teaching.

Furthermore, many scientists and engineers believe that the act of networking is a smarmy enterprise, relegated to the purview of a used car salesman. Some of my fellow nerds incorrectly refer to networking as "schmoozing" and think that it entails bragging and other potentially sleazy actions that endeavor to get something from the other party and almost take advantage of them. But as I discuss in Chapter 1, networking is the exact opposite of this. Networking entails providing authentic and genuine information for and between both parties so that you both can contribute value to each other's projects and interests. It is about building long-term, win-win collaborations, not desperately trying to dupe and fleece the other person. As nerds, we are far classier than that!

And I want to be clear – I am a nerd. I have a bachelor's degree in mathematics, where I focused on extremely theoretical stuff – the formulae that describe donut holes and knot theory, for example. I also studied physics and a little astronomy and was President of the University of Arizona Society of Physics Students (a position I got, and a position I leveraged to get other opportunities, through networking). I enjoyed studying for math tests, watched marathons of Star Trek, hung out with other nerds making physics jokes, and spent summer Saturdays going to lectures at the Princeton Plasma Physics Laboratory while I was growing up in nearby Princeton Junction, NJ. These nerdly tendencies have continued as I transitioned to adulthood. I play math puzzles for fun, incorporate Klingon vocabulary into my everyday speech, read (and write for) science and engineering magazines with a voracious appetite. My friends are all nerds. I proudly proclaim "I am a nerd!"

So I have extremely great respect for other nerds, especially those who endeavor to improve their skill sets to ensure they can achieve their nerdly dreams, whether they are professional, scholarly, or even personal in nature.

I wrote this book because nowhere in my mathematics education did I ever have a professor or advisor tell me how to network or that I should even be doing so. It was an extremely foreign concept. And yet, for me, when I actually began to network, many of the concepts described in this book felt intuitive to me. Additionally, I realized that I had observed others finding professional success and bliss through networking. I knew I had to share these ideas and inspiration with my fellow nerds.

So this guide is designed to teach you why networking is essential, how you can utilize networking as a strategic tool in your career planning and job searching, and to dispel any negative myths about networking. You'll gain tactics and strategies to build diverse networks, find people for your networks, and access new networks for new career opportunities. You will gain knowledge about how to organize and maintain you networks. You will learn how to network at an event and how to effectively use social media channels to expand your networks. And finally you will discover how to preserve the networking momentum you generate with other parties, no matter where life takes you. Remember: Networking is a gift that keeps on giving and follows you from job to job, organization to organization, and career to career.

As you read this book, I hope you will also recognize how much power you have, as a nerd, as a scientist or engineer, and as someone who endeavors to connect with others. In many respects you are an entrepreneur – the CEO of your own company – "Me, Inc." You are in charge of all aspects of advancing Me, Inc. – not the least of which involves being innovative and entrepreneurial in your networking activities. In fact, you will see that you can even create your own networking opportunities or "Networking Nodes," which I define as any thing, person, place, conference, LinkedIn group, Twitter feed, and so on that draws like-minded individuals together and is thus perfect for networking. Think entrepreneurially, think innovatively about your networking and your career triumph will follow in parallel.

But of course, no matter how much time and energy and focus you invest in networking, you must never lose sight of your outputs – your productivity in your field and profession must be sustained at high levels in order to remain and sustain success. Your stellar brand (promise of value) and your reputation, which serve as fuel for effective networking, must never be allowed to incur damage through non-productivity. I write more about this in Chapter 1, but it is an important notion to always keep in mind: You can't expect to be successful in networking if you are not successful in your profession and job and vice versa. So this book is meant to give you specific steps to take to incorporate active and passive networking into your career plan and even day-to-day activities without sacrificing both the quality and quantity of your vocational outputs.

The book is also crafted to give you, my nerdly brethren, many of whom are introverts, a boost of confidence in your networking abilities. You don't have to be afraid to go up to someone and introduce yourself at a conference. You can walk into a reception and, not knowing anyone there, lay the foundation of a fruitful alliance. Networking is an achievable (and even fun!) enterprise which will only make your life all the more rich, and I will show you how to do this.

And speaking of riches, I couldn't have gotten this far in my career(s), or furthered my own networking goals, had it not been for some very important people in my life, all of whom I met through networking. I wish to thank them with all my heart:

Daniel L. Stein, who, as head of the University of Arizona Physics Department, was my first boss out of college when he hired me as his Director of Communications. He knew me as a student and as a leader in the Society of Physics Students, but he took a chance on me when he could have hired someone with experience and better qualifications. Somehow he knew I would provide value. His seemingly simple decision changed the course of my life for the better, and I am so grateful for the time I had under his leadership and for the mentorship, guidance, and friendship he has shown me throughout the years since. Even after he left the UA in the mid-2000s and became Dean of the College of Science at New York University, we stayed in touch. Both he and his wonderful wife, Bernadette, have been great sources of strength and direction for me for years.

Alan Chodos, the former Associate Executive Officer of the American Physical Society (APS), whom I met more than 14 years ago, as a direct result of networking. In 2001, I attended an APS conference and since I had just transitioned from the job as the director of communications in the University of Arizona Physics Department, I asked an organizational representative if I could hang out in the press room at the meeting. I was gladly offered this chance, where I met the APS director of public relations. Four months later, this director emailed me and asked me if I was interested in his job, as he had just left for a new position at the National Academies. I said I was and started corresponding with the person who would have been my boss, Alan. In the end I declined the job. But I stayed in touch with Alan and worked on and off for him on short-term projects for years, before suggesting I write a column for him in 2007. Since then he served as my editor for Profiles in Versatility, a column in APS News, concerning physicists in non-traditional careers, until his retirement in 2014. Alan continues to be a fantastic mentor, guide, collaborator, and pal, and I am so appreciative of his support of my career. And if it hadn't been for networking I may never have met him.

Jerzy Rozenblit, a colleague, client, and friend with whom I first became associated thanks to a mutual colleague who introduced us. As Chair of the Department of Electrical and Computer Engineering at the UA, Jerzy was looking for someone to organize a gala fundraising event celebrating the 100tb anniversary of his department. A member of his staff knew me and knew my work and recommended he speak with me. This initiated a fruitful alliance and gave me the opportunity to work in a completely new realm of science communications. Most importantly, I got to know Jerzy and his team, and we both discovered how much we enjoy working together. Thank goodness for networking and reputation management activities!

Joaquin Ruiz, my second boss, the dean of the College of Science at the UA, who gave me great leadership and offered me amazing opportunities to grow and advance and learn. Working for Joaquin was like getting an MBA – I gained exceptional business skills, and he instilled in me a desire to push myself to always deliver the best. I certainly would not be as successful as I am today had it not been for him and his trust. And by the way, I got the job with Joaquin because he knew me by my reputation!

Sarah Andrus, who helped me fine-tune my book idea for publication and whose friendship I cherish.

Justin Jeffryes, my editor at Wiley, whose guidance, direction, and patience helped make this book a success.

There are yottatons of other people who have helped me in myriad ways throughout the years, and whose assistance I have appreciated in the process of this book. These are the people who have opened doors, introduced me to colleagues, created ways for me to inject value in their organizations and designed new opportunities for me to advance in my career. To all of my friends, compatriots, and teammates who have been on this journey with me, I affirm my utmost gratitude to you.

And to you, the reader, my nerdly ally, I thank you for trusting me with your networking and career advancement goals. Here's to your professional success, personal bliss and all the profits that will bring you and your partners through effective networking. Enjoy your networking!

1 The Importance of Networking and the Hidden Platter of Opportunities™

Networking is a necessity, but what exactly is networking? In this chapter, we introduce the concept of networking and how you can apply it to gain access to hidden opportunities.

A few years ago, a freshly minted science doctorate asked me for some help finding a job. He had applied for hundreds of advertised openings, both postdoc and non-academic positions, but to no avail. So I asked him about his networking strategy. "What networking strategy?" he replied, clueless as to what I was referring. I spent the next hour emphasizing the importance of networking in finding hidden job opportunities and communicating your value to decision-makers. I outlined for him a customized networking plan which would enable him to meet and interact with professionals who have the power to hire him for the jobs he so desperately wanted. When our meeting concluded, I asked for feedback on the career consulting session – "Did you find our discussion helpful?" I inquired, thinking I was up for significant praise. "No," he said instantly. "You didn't tell me where I can apply for a job or places where there are more advertisements for jobs."[i]

Although he has a PhD in chemistry, this scientist did not understand a fundamental element associated with career planning and job seeking: Most jobs are NOT advertised, and neither are most opportunities that have the potential to be career game-changers, such as invitations to meet with someone, serve on a committee, pursue a leadership role, or apply for an award. And for career opportunities that are promoted, like postdoc appointments and academic professorships, often times, the committees already have people in mind whom they want to invite to apply or have promised the job to someone under the table.

How do you position yourself so that you can find out about the hidden jobs and other opportunities and be considered before the rest of the herd? It's all about networking. It's *all* about getting your name and accomplishments out there, managing your research reputation, and connecting with as many influencers and decision-makers as possible. And although many scientists do understand that some sort of networking is needed, they don't often understand what it constitutes and why it is obligatory in gaining a competitive edge. Some scientists, like the chemist I was counseling, erroneously think that networking is a side activity that won't lead to anything solid, like a job. But that's simply not true. The sooner you recognize that networking is actually a strategic tool in finding a job, defining a career path, and even advancing scholarship itself, the sooner you will set yourself up for success.[ii]

In fact, networking is the most powerful tool you have in your career planning kit. It is the secret to finding hidden, game-changing opportunities; it establishes and solidifies your unique value in the minds of decision-makers; it opens doors to people, places, alliances, and information that you didn't even know existed. The bottom line – networking gets you jobs and considerably contributes to your career advancement.

But the challenge of networking is that most people don't exactly understand what it is or means, nor how to start.

Some people think that "networking" is a single, finite action that takes place at an event, like a conference. They meet a professional at a mixer, they chat, they exchange a witty quip, they enjoy the meatballs and other finger foods. When their 10 minutes is up, or they can't think of anything else to discuss, they excuse themselves and voilà! The networking is complete. They pat themselves on the back for a job well done and because they have done their requisite "networking" for the week (or year). They don't follow up with the person; in fact, in many cases, they never speak with them again. But that's ok because they have "networked."

What is the outcome of their celebratory "networking"? Absolutely nothing – no new information, no new career opportunity, and certainly no new potential collaboration. They may wonder why their networking didn't help them achieve anything. And they go back to applying for jobs advertised on the internet and banging their head against the wall.

The truth is that networking is not a one-time deal, but rather something much, much more. Here's the breakdown. Networking is:

- a spectrum of activities …
- which begins with that first interaction …
- and continues throughout the life of both parties.
- It aims for a mutually beneficial, win-win partnership …
- and involves myriad correspondence and actions …
- that provide value to each party …
- and only ends when one or both of you drop dead.

Yes, it is just that simple.

But I want to point out one component of the definition of networking that I give above. The idea of networking, or a networked partnership, continuing throughout the life of both parties may seem a little intimidating to some. I completely understand that sentiment. When you realize that the relationship you are launching with another party will last both of your lives, it can almost appear overwhelming and could prevent one from initiating contact with another. But let's look at this from another point of view: Networking is a gift that keeps on giving. Once you introduce yourself to someone, you are now networked with them. You can maintain and carry your connections made through networking throughout your life, and from job to job, from career to career, and from location to location, to the benefit of both you and the other person. Just because you change industries or move to a new geographic region, doesn't mean you lose your networks or the partnerships you have built from networking. You can and should continue to nourish your networks wherever life takes you with the knowledge that networking, when done appropriately as I describe here, will continue to deliver different packages of value at different times in the relationship. So it can last a lifetime, because once you are connected to someone you can always reach out to them again. I have contacts whom I contact maybe once every five years. But this is fine, because that's the nature of our particular partnership. At another time, maybe in another 10 years, I will reach out to them again and perhaps we will start a specific project together. But the point is that opening is always there for me, and vice versa, until we both shuffle off this mortal coil. So don't let the possibility of a fruitful, mutually-beneficial lifetime relationship with another person stop you from networking.

Once you recognize these basic principles of networking, you can begin to craft your own networking strategy to help you achieve your career ambitions.

But before you do so, I want to clarify how networking success not only helps you, but also greatly benefits the entire science, technology, engineering, and math (STEM) community. In fact, even if you don't care about your career at all, but still love science and engineering, you need to engage in strategic networking if for no other reason than to advance our understanding of our universe. The reason is clear: Science and engineering research does not exist in a vacuum. It requires diversity among groups of dedicated professionals who constantly are inspired to find novel approaches to scientific problem solving. You cannot have innovation without a regular influx of a diversity of ideas, which requires a regular influx of partnerships with people who have diverse educational, disciplinary, and cultural backgrounds. And to find and access people with whom you can craft collaborations and thus improve your productivity, you have to network. Networking is the ticket to the next scientific revelation, the next engineering breakthrough, the next "big thing" that contributes to humanity's wellbeing and illumination of nature's truths.

But of course you care about your career – you wouldn't be reading this book if you didn't. This book will help you develop and implement a networking plan that mirrors your professional goals, contributes to your discipline, and gives you access to hidden career opportunities.

Eight Networking Myths

Let's start by analyzing a few networking myths:

Myth Number 1: "I don't need to network, because my excellence in my field alone will ensure I advance in my career"

Now I know what some of you may be thinking. As a scientist and engineer, you don't have to network. Your "outputs" as a technical professional, either in the form of research, published papers, presentations, mentoring and supervision of protégés, teaching, patents, and/or any other activities that contribute to the scholarship of science and engineering, are all that is required for you to get a job. Decision-makers will read your curriculum vitae or résumé, realize that you are the answer to their professional needs, and hire you because you have proven your abilities through your outputs. I speak to young scientists and engineers about this all the time and hear the same complaints about networking – it is not needed, it is a time drain, it is a distraction from my technical outputs which will really serve to get me the next job.

And in fact, your advisor might even discourage you from networking, because they themselves don't understand its true definition. But know this: They didn't get to be the principal investigator of a research group or the head of an engineering department because they rested on their own laurels. They achieved career advancement because they engaged in networking. They may not have called it that. They almost certainly didn't think of it as self-promotion, marketing or branding, all completely legitimate concepts (closely affiliated with networking) which I discuss in this book. But they did go to conferences and speak with other professionals, they did read research papers and reached out to the authors to discuss ideas for collaboration or to share information, and they did attend subject-based meetings which resulted in new problem-solving methodologies and new alliances, all of which begat new career opportunities. And all of these activities fall under the umbrella of "networking." So even if your mentor doesn't refer to it as such, you have to network and, in doing so, articulate your value to others if you want to advance in your career. You can't expect your superior abilities to sell themselves – you have to tell people about what you do and the value you can provide them so they understand how an alliance could be mutually beneficial.

Myth Number 2: "Networking with people outside of my field is a waste of time"

I read someone's brilliant advice along these lines on LinkedIn recently. The gentleman encouraged young scientists to interact only with other scientists, and he rationalized this gem by indicating that a plumber, for example, could never help you in your career. But of course this is faulty thinking. You want to interact with people in and out of your field, industry, and even geography as much as possible because anyone could be the person who provides you with ideas, information, or inspiration that could take your career and scholarship to the next level. They don't have to be in science to help you solve a science problem. They don't have to be in academia to know someone who can help you land a job in higher education.

This kind of strategic thinking has aided me many times. Once, while sitting in the middle seat of an airplane, I overheard the two people on either side of me discussing their plans for interacting with editors. My ears perked up and I interrupted their conversation to introduce myself. "Are you in journalism?" I inquired, getting excited because I am a science writer. No, they responded. They were in public policy for the nuclear regulatory industry. Now, at the time, not only did I have no interest in pursuing a job in nuclear science, the sector wasn't even on my radar as a potential ecosystem in which I could contribute or collaborate with others. I never would have even thought about it as a career choice. But once I got them talking about their passion for the field and what exactly they did, all three of us realized there may be an opportunity to partner on a project. They told me about a major nuclear science conference that occurs in my state every year, and I was able to convert that tip into a real work opportunity.

You just have to start a conversation to learn what value you both could provide. So think big and aim to engage people from all walks of life. And remember the theory of six degrees of separation – we are connected to everyone else on the planet by no more than six degrees, so the more people you know, the more people you have access to.

And just to clarify: Your goal is not only to connect with as many people as possible, but to build as many networks as possible too. Networks may be defined by many descriptors such as location, discipline, sector, culture, nationality or language. They can feature people in science and engineering and in insurance, consulting, surfing and landscaping. They can include people who are currently employed, underemployed, or unemployed, and they can be organized by hobbies, skill sets, aspirations and lifestyle choices. But they all have one characteristic in common: Each network contains many people all of whom have the potential to exchange ideas and inspiration with you to propel your career (and even life) into a new realm. So seek to engage and formulate multiple, diverse networks to ensure you have a continuous stream of inspiration to solve your problems.

Myth Number 3: "Networking is about extracting something from someone else"

The foundation of networking is building strong collaborations which provide value to both individuals. Of course you need a job, but the other guy needs things too, like connections with other potential partners, career leads, or ideas about funding sources. You should seek to craft an alliance that is not about grabbing what you can from each other, but rather learning what each of you can give to ensure that the relationship continues to harness and exchange both of your specific values. So explore each other's interests and particularly for early-career professionals, offer to be of assistance even if you don't immediately see a potential return on investment (ROI). You'd be surprised how much others appreciate your proposal to aid them in some way and how that can solidify the relationship from the get-go. I once read an article in a trade magazine that I especially enjoyed and contacted the author to discuss it. She was more than happy to chat with me about her passion for the subject that drove her to write the essay. At the end of our conversation, I informed her that if I can ever help her in any way to please contact me. "Even if I am not the right person, I will find the right person or resource to aid you in your quest," I stated. She expressed gratitude and followed up with me in a few months. That one gesture started a partnership that has lasted more than a decade, and led to the expansion of more networks for both of us. These types of partnerships are gifts that keep on giving – to everyone involved.

Myth Number 4: "I don't need to network because I don't need a job now"

Networking opens your mind to new possibilities for careers and jobs. But it also works in the other direction as well – by networking, you make yourself known to decision-makers who may want to engage you now or in the future. Networking, as mentioned above and detailed below, gives you access to hidden career opportunities because networking is inherently linked to self-promotion. This is a very important concept – when you network, when you speak with someone about your expertise, experience, passion, and talents, you are promoting yourself. There is nothing wrong with this enterprise – after all, how could you even begin to network if you didn't introduce yourself to others and share your expertise with them? When you do so, you are essentially promoting what value you could potentially provide the other party. This is very useful information for the other party to have, as it helps them start analyzing ways in which you both might be able to collaborate. This may lead to a detailed conversation now, or the chance to serve on a committee, or to co-author a grant in the near future. And these activities can lead to a job later. So don't wait until the last minute to start your hard-core networking or to execute a networking strategy – networking is a long-term endeavor that may involve many conversations and exchanges of value before a job is proffered.

> TIP: Networking is a gift that keeps on giving.

Myth Number 5: "I can't network because I don't have time"

There is a tragi-comical paradox that is associated with networking and career progression: Networking is practically a full-time job in and of itself, which you have to pursue while you also have the full-time job of career development and your other full-time job of being a scientist or engineer. So basically you have three full-time jobs and yet there are no parallel universes or time machines at your disposal. While it is true that networking does take an abundance of time, it is time well spent: I can almost guarantee that resources you devote to networking will provide a substantial ROI. In fact, you may even get more out of it than you estimated.

A smart way to get started is to recognize that there are two kinds of networking: active and passive. Active networking is where you purposefully seek out others to meet and with whom to connect. This may occur at conferences and symposia, institutional and departmental events such as colloquia and journal clubs, or through reading papers and contacting authors. Passive networking entails interacting with others who happen to cross your path. This could happen at affairs that you are attending which are not related to science, such as philanthropic or pastime activities in your community. My favorite kind of passive networking occurs on airplanes. You are trapped next to the guy who is less than 3 nanometers away from you for a big chunk of time, so you might as well make the best of it. As long as he's not drunk, strike up a conversation and you never know where it could lead. As illustrated above, just from passively networking with people on planes, I have gained surprising knowledge about my industry, developed partnerships with new colleagues, and even landed a few gigs.

And don't underestimate the use of social networks like LinkedIn for both active and passive networking. More and more, LinkedIn is becoming a standard for hiring people – in fact, one industry decision-maker

TIP: A LinkedIn presence is a requirement in today's networking. You must be seen in this professional marketplace.

told me that the résumé is becoming obsolete and the LinkedIn profile is taking its place. You should create a free profile on LinkedIn and join groups that are relevant to your ambitions. I will go into details about utilizing social media channels for networking in Chapter 8.

But the most important point to remember is this: "Successful" people – professionals in career paths that bring them delight, excitement and joy, provide exciting challenges and allow for the utilization of high levels of skill and creativity – are in those positions because they networked. Moreover, they didn't stop networking once they reached a career pin-

TIP: Successful people do what unsuccessful people are not willing to do.

nacle; in fact they network all the time. Successful people don't stop networking, because they know that the bigger and more diverse their networks are, the better they will be at their profession. Whether subconsciously or consciously, they look at every opportunity to speak with someone, whether they are attending a conference, on an airplane or in the grocery store, as a chance to network and grow their knowledge base and solve problems in new ways.

Myth Number 6: "I can't network effectively because I am shy and introverted"

Efficient networking, where you are able to build long-term, win-win partnerships, takes practice. You don't have to be an extrovert to network, and even outgoing personalities (and seasoned networkers) sometimes have butterflies in their stomachs when they first approach a stranger at a mixer. But the more practice you get at introducing yourself to others, the more comfortable you will get, the easier it will be and the more astute you will become at networking.

But I want to be clear that being a successful networker does take courage. Most people THINK they lack the necessary courage because they don't know the means by which to actually go about networking, nor do they understand the ROI they can get from networking. But I want to emphasize that I know, deep down inside, you do have this courage. How do I know? Because you chose to study and pursue a career in the most difficult and potentially scary subjects on Earth – science and engineering. Very few people have the nerve to approach science, let alone devote years of their life trying to solve problems which either have never been solved before, or need to be solved in novel ways. Your decision to pursue the unknown landscape of STEM demonstrates an innate nature that is clearly fearless: So if you are brave enough to chase science and engineering as a career, then you are certainly brave enough to start networking strategically. And you can leverage this boldness to buoy your confidence when networking.

At networking affairs, when I approach someone and after I introduce myself, one of my favorite opening lines is "what's the best part of your job?" This inviting inquiry reminds the other person about what brings them bliss. And as they start to recount what is pleasurable about their work, they will be more apt to speak with you about it. One of the secrets of networking is that people generally love speaking about themselves. So the more you ask them about themselves and what drives them, the more you are able to start the relationship off on the right foot. As they speak, remain in eye contact. I like to jot down a few notes as people chat on the back of their business cards. So when I follow up later, I remember (and can remind them) of what we addressed in our conversation. (I will go into more detail about networking at events in Chapter 7.)[iii]

Myth Number 7: "Networking is a smarmy endeavor relegated to the domain of a used car salesman"

In certain circles, "networking" has been given a bad rap. Some people think that the act of networking, like at an event or even online, can seem sleazy and unauthentic, like a used car salesman trying to unload a lemon. And while I am sure that there are certain people who do perceive networking to have this air of negativity, they are usually the same people who think that networking is only about extracting something from the other party. In actuality, if you network appropriately, you are giving the other party information which can assist them in making a decision that is designed to benefit you both. When you network, you are not selling a crappy car, you are articulating your goals and expertise and passion in an effort to discover how you and the other party might craft a win-win partnership over time. There is nothing sleazy about two people exploring avenues for collaboration; on the contrary, it is always a privilege to have the opportunity to share ideas that could spark an alliance. So it is important for you to be authentic and honest about your skills and the value that you can provide.

Myth Number 8: "A leader in my field would never want to speak with me, an early-career professional"

An early-career astronomer once told me that there was no point in going to a reception at a conference, because the "stars" of the astronomy world, a.k.a. the observatory and department heads, program managers, and international leaders, are not interested in chatting with a "lowly" grad student. "They don't want to speak with me," she informed me. "I'd be wasting their time." This is a myth! At a networking event, like a reception or mixer, especially one at a conference, established scientists and engineers want to meet each other, AND they want to meet the emerging professionals as well. After all, the grad students and the postdocs are the future stars of the field. They are the scientists and engineers who will add energy to a research program, and continue investigations into novel directions, now and in the future. In fact, many senior-level leaders consciously recognize that early-career professionals will serve as the legacy of their own work down the road. They need you, just as much as you need them, and this is the essence of the win-win element of networking. Remember: No matter where you are in your career, you always have something of value to share, even

> **TIP:** No matter where you are in your career, you always have something of value to share, even if it is with someone who is in a seemingly "higher" position than you.

if it is with someone who is in a seemingly "higher" position than you. At your next networking function, don't hesitate to walk up to a keynote speaker, journal editor, or principal investigator and ask them about their work. Talk to them about your interests and ask them about their passions. Discuss ways in which you may be able to contribute value and collaborate on a project.[iv]

Now that you recognize a few of the fundamentals associated with networking, let's address one of the best and also least-understood benefits associated with networking: Accessing hidden career opportunities.

Understanding the Hidden Career Market

I can't overstress the significance, power, and extent of hidden career opportunities in an overall career plan, no matter what industry you desire to be in. Understanding, accessing, assessing, and ultimately harnessing hidden career opportunities is THE gateway to surprising professional benefits and advancement for both you and your collaborators along your route, and you must constantly be on alert for them.

And the key to unlocking them all is through networking.

But before you start looking in every nook and cranny for hidden jobs and other advancement opportunities, there are a few basic aspects of the market for hidden career opportunities that you must comprehend. First and foremost, know that hidden, game-changing career opportunities are everywhere, and they come in many forms. A hidden career opportunity could be as direct as an invitation to apply for a job, or something that requires more cultivation, such as the chance to collaborate on a short-term project, serve on a committee, or simply have a conversation with someone.

Don't think that an opportunity to have a cup of coffee with someone is any less valuable than an offer of employment itself. Rather, the chance to engage this person and discuss your mutual interests will help you to craft a strong partnership. And this alliance can and does lead to actual jobs.

Sometimes an opportunity that appears not to be concealed, such as an open, advertised job on a company or university website, is in actuality hidden. Many jobs are promised to candidates "under the table," meaning they are invited to apply and offered the opportunity, but due to legal or other restraints, the organization has to advertise the position anyway. You might notice this in cases where a job is advertised and then the ad is removed within a week. Did the organization really find and hire a qualified applicant within seven days? Although it has been known to happen, it is not the norm. More likely, the person who got the job found out about the opening and ultimately landed it via the hidden market.

Breaking into the Hidden Job Market

So now that you know a little bit about the hidden market for career opportunities, allow me to present some principles for piercing and leveraging this arena for your own professional prosperity:

- Don't try to quantify it: I know you want to. I understand the urge (after all I studied math too) to attach a number to the hidden job market so that you can very carefully develop a statistics-based approach for pursuing and applying for jobs. But depending on which career expert you consult, each person will probably give you a different number as to how much of the job market is clandestine – it could be anything from 40–95% of jobs and other career opportunities. I personally believe the right number hovers around 90%, based on my own experiences and other factors (see below). Instead of spending valuable time trying to analyze exactly how much the hidden job market encompasses, I recommend looking at it as a binary issue. Recognize that it simply exists. It is a part of the job market, and if you are to advance you have to cover all avenues for finding and landing jobs. Therefore you have to look for hidden opportunities. It IS a black and white issue.
- It is accessed only through networking and reputation management activities: If you want to find out about hidden career opportunities, you must make yourself and your brand (promise of value) known in your community or industry. Networking is the most powerful way to do this. It is designed to build win-win relationships between parties, and the more you know about each other, the more you will realize what hidden opportunities exist that can serve both of you. For example, you might meet someone at a conference and ask them out to lunch, and while chatting with them over tuna fish, he learns that you speak Spanish fluently. It turns out that he has a project in

Buenos Aires and he is looking for someone with the technical talent and linguistic acuity that you possess. You have now uncovered a hidden career opportunity that you might never have known about had you not sought to network with this professional. As people get to know you, they will begin to present you with previously concealed opportunities – I call it the Hidden Platter of Opportunities, because it literally is offered to you. It happens like this every day.

- Any opportunity can be massaged into a networking opportunity: This is yet another great example of how networking and opportunity procurement cycle back into each other. Via networking, I may unlock information pertaining to an open spot on a board of directors. I take advantage of this opportunity to serve on the board, which grants me the opportunity to network with the other members of the board and people in their networks. And on it goes and grows.

- You contribute to it too: Just as other scientists and engineers have access to information, ideas, people, collaborations, and actual jobs, you do too. You just may not realize it. But since the core of networking is providing an exchange of value between parties, you can provide access to hidden career opportunities for the people in your network. Doing so will help establish your reputation as a thought leader in your field and will encourage others to want to network with you. I noticed this result recently after I learned about a number of fellowships for scientists and science writers, two of which included a $10 000 prize. I perused my groups on social media to see if anyone was promoting these and was surprised that others had not heard about them. (I had only accidently discovered them myself while surfing the web.) So I shared them on LinkedIn, Twitter, and Facebook, and the responses I got were extremely positive: People expressed gratitude to me for communicating these hidden opportunities and as a result I was able to connect with people I might not have had access to otherwise, and demonstrate my commitment to my community, thus solidifying my brand. I am sure to learn of other opportunities, which will benefit both me and my colleagues, as a result of this one action.

- Most people do not pursue opportunities: As I noted above, as a scientist and engineer, you have an advantage that other people may not have, in that you possess an element of fearlessness. It may be secluded, deep in your soul, but it is there – you wouldn't have pursued this profession if you didn't have it. This bravery will aid you greatly throughout your career and especially when it comes to pursuing the Hidden Platter of Opportunities. Most people, no matter their vocation, do not take advantage of opportunities, whether they are clandestine or not. They don't ask other professionals out for coffee, they don't send "cold emails" (correspondence with those they don't know), they don't apply for awards, and sometimes they don't even apply for jobs even when an advertisement is staring them in the face.

The main reason for this is because of fear. They may be afraid that the outcome will be negative. If they ask for help, the answer will be no, if they apply for a fellowship or submit a book proposal they will be rejected. Of course there are other cases, in which some people don't chase opportunities because they are afraid the outcome will be positive. "What if I apply for the job and get it? What will I do then?" They fear success in part because they don't think they can live up to their reputation or compete against their peers.

Since the majority of people don't hunt and take advantage of opportunities, this gives you a distinct competitive gain. By simply asking for a meeting, discussion or other opportunity, you are articulating your brand to the other party and exhibiting the confidence that you can help them achieve their goals. This in turn gives you access to even more hidden opportunities. Do not be afraid to pursue success!

- Every opportunity leads to another: One of the best elements of taking advantage of opportunities is that you can almost be assured that it will lead to another, often better opportunity. I have seen this myself throughout my own career. For example, many years ago, I volunteered to serve on a committee, which led to be being elected the president of the committee, which led to an invitation to apply for a job, which eventually led to me getting the job. Yes, it can be that straightforward!
- It can also involve you creating your own opportunities: Don't forget that the ultimate hidden career opportunity, the one that may bring you the greatest return on your investment of time and energy, is the one you create yourself. Remember, Bill Gates didn't apply for an advertised job – he made one himself and launched an industry. You should always be thinking entrepreneurially. If you need an opportunity ask for it. If it doesn't exist, create it yourself. You may just start a revolution.[v]

> **TIP:** If you need an opportunity ask for it. If it doesn't exist, create it yourself. You may just start a revolution.

Every day I am amazed at how simple actions that don't even seem like "networking" manifest themselves and allow me to access hidden opportunities. While simply browsing through LinkedIn for example, one day when I was bored, I found numerous alumni associated with my alma mater, who have similar backgrounds and interests as I, who work for organizations that I would like to engage, and whom I did not know personally. I immediately became excited because I could see how partnerships between myself and the other people could be developed that would benefit us both. This was a passive action I was taking and yet it created many, many potential hidden opportunities to contact these people, which ultimately could reveal its own hidden opportunities for shared projects. And that's one of the beauties of networking: The more you do it, the easier it becomes and the more you do it without even realizing you are doing it. It becomes second nature to you. And that is one of my goals for writing this book in the first place. And now, on to networking!

Chapter Takeaways

- Networking is a necessity for your own career advancement.
- Networking is a requirement to influence the progression of scholarship.
- Networking is not a one-time act – it is a spectrum of activities that aims for a long-term, strategic partnership.
- Networking is inherently linked to self-promotion: Communicating your interests, value, and expertise.
- Networking gives you access to the Hidden Platter of Opportunities.
- Game-changing career opportunities are everywhere and can lead to other opportunities.
- Most people don't have the guts to pursue opportunities, which gives you a distinct advantage if you do.
- Take advantage of as many opportunities as possible, and if you need an opportunity ask for it. If it doesn't exist, create it yourself. You may just start a revolution.
- The more you network, the more it becomes second nature, and thus the more you do it.

Notes

i. From Scientists Can't Network and Other Myths, *Euroscientist*, March 9, 2012, http://euroscientist.com/2012/03/scientists-cant-network-and-other-myths/.
ii. Ibid.
iii. From Networking: It's More than Sharing Meatballs, *Physics Today*, April 3, 2013 http://www.physicstoday.org/daily_edition/singularities/networking_it_s_more_than_sharing_meatballs.
iv. Ibid.
v. From Understanding Hidden Career Opportunities, *Physics Today*, July 10, 2013 (not completely verbatim) http://www.physicstoday.org/daily_edition/singularities/understanding_hidden_career_opportunities.

2 Understanding and Articulating Your Value Proposition

In order to network successfully, you have to first identify and then be able to communicate your value.

You are a lot more important than you might have thought

Every product, service, organization, company, and entity on Earth has a scientific discovery and/or an engineering innovation at its core. Think about it – can you imagine anything that did not require scientific discovery or problem-solving to uncover its natural secrets, or engineering innovation to bring it to existence? No matter that *it* is, science and engineering had to make *it* happen. It could be something as seemingly simple as a notebook or as complex as a city. Science and engineering birthed it and probably reared it as well.

> **TIP:** Since everything on Earth has a scientific discovery and/or an engineering innovation at its core, this means you have a lot more value (and power to apply that value in myriad careers) than you probably even realized.

And since science and engineering serve as the foundation for everything, and you are the propagators of scientific and engineering knowledge and innovations, that means that you have a lot more power than you realize, especially when it comes to career planning and advancement. As the knowledge generators and the value creators, you are in a unique position, which grants you seemingly infinite career choice. This is due to both the value you create and discover, and the value you gained pursuing science and engineering, on which I go into detail below.

Wow! In some respect the universe really does revolve around you.

Not everyone sees this value proposition. I once had an advisor outline available career options for someone, like me, with a bachelors in mathematics. Other than my becoming a professor or teacher, or going into actuarial studies, he intimated there was nothing I could do with my math

Networking for Nerds: Find, Access and Land Hidden Game-Changing Career Opportunities Everywhere, First Edition. Alaina G. Levine.
© 2015 John Wiley & Sons, Inc. Published 2015 by John Wiley & Sons, Inc.

degree. Nothing – as in zero career opportunities. He advised me to go to grad school and give in to academia.

At the time I was shocked and dismayed, but in the end I ignored his advice. Instead I launched into science communications and have since crafted an intellectually stimulating career, anchored by my love of science, at the crossroads of writing, professional speaking, career consulting, comedy, and even event planning (so if you have a wedding that needs to be arranged, please tweet me). I consider myself lucky: I was able to figure out that people with science, technology, engineering, and mathematics (STEM) degrees have n career options, where n is significantly greater than zero and, in fact, theoretically approaches infinity.

So how do you find these careers, access them, and assess whether they are right for you? Networking will be the channel, but first you have to understand how much value you have as a professional with an education in STEM, whether you are just graduating, completing your postdoc, or have been working for 10–15 years or more. Most scientists and engineers think their only value lies in the subject of their expertise: I am a physicist, therefore I can only do physics. And although it is absolutely true that you have great talent in conducting scientific or engineering research, you have much more to offer potential employers. You have *highly coveted skills* that you gained simply as a byproduct of studying STEM fields. As a result of your schooling, you are:

- An exceptional problem-solver who can see and solve puzzles both granulistically and holistically and in many dimensions.
- An amazing critical thinker who can analyze and imagine situations and scenarios with a 360-degree perspective.
- A superior researcher, with the ability to find answers to complex and seemingly impossible questions.
- A talented project manager who can multi-task on diverse teams with diplomacy and great aplomb.
- An adaptive and flexible hard worker.
- A person who is knowledgeable about how the physical world works, and knows how to apply that knowledge to solve problems in novel and everyday realms.
- An extremely disciplined, self-motivated, self-reliant professional (you have to be if you decided to and then successfully researched and pursued a STEM field for your vocation).
- Brilliant, brave, and a risk-taker, by virtue of the fact that you chose to pursue a STEM discipline and succeeded in doing so.

Your Unique Problem-Solving Abilities: The Cornerstone of Your Value

Problem-solving is listed first here because it is the most important skill anyone has. In fact, when you are hired for any job, as a professor, a

> **TIP:** The purpose of every job, in every organization, in every sector, in every part of the known universe, is to solve problems.

president, or a custodian, you are hired to solve problems. And if there is one thing scientists and engineers excel in, it is solving problems. This is not to be taken lightly. Most people don't realize that their foremost skill of importance is problem-solving, and as a result they don't articulate that to potential collaborators and employers. They don't discuss their problem-solving agility in detail on their CVs, they don't verbally communicate it when they are at a networking mixer and they certainly don't clarify it during a job interview.

But if you can clearly enunciate your value, in part as a function of your unique problem-solving experience, expertise, and savvy, you will unlock hidden career opportunities and you will be more able to access advertised openings. This is the essence of effective networking.

This task of precisely expressing your value is not that difficult to do (especially once you realize the extent of that value). And when you start doing it strategically (or even passively), you will find that hidden opportunities will pop up in surprising places. Years ago I was flying from Dallas to DC and as I was boarding the airplane I noticed my seatmate's shoes. They were just so beautiful and one-of-a-kind that they easily stood out. As I sat down, I couldn't help but compliment this stranger on her colorful choice of footware. She smiled and I soon introduced myself, and before long we were just chatting away at 30 000 feet. We both became so immersed in conversation that by the time we landed, two hours later, we could hardly have guessed the time went by.

As it turns out this woman was the wife of a very well-known senator and as we conversed she asked me about my work. I told her that I was a professional speaker and comedian. She seemed engrossed by this and asked me a lot of questions about it. She soon revealed that she does charity work which requires an abundance of public speaking and is driven to enhance her speeches. But injecting them with humor was a challenge – "do you think you could help me with that?" she inquired. And of course I said yes, and gave her some tips right then and there. I got her business card and gave her mine, and later sent her a thank you note. I also emailed her to stay connected and offered to assist her further with her speechwriting needs.

This is a classic example of my point – if you can plainly articulate that you can solve the other person's problems, this will entice them to want to learn more from you, which will most likely result in them offering you access to the Hidden Platter of Opportunities. Perhaps you will make the opportunity yourself or the other party will customize one just for you, as the senator's wife did for me. Or maybe in the course of the conversation the person will realize that you can solve certain problems in x or y fields or departments within their organization, and they press you for more information that ultimately helps them make a decision about whether to engage you further. Just remember – everyone has problems that need solutions. By networking strategically, you can begin to communicate that you can provide those valuable solutions.

One of my favorite tales of the merits of articulating your own value while networking features a professor of physics. Alexander Skutlartz was a particle physicist at a university in North Carolina and was married to an expert in grain science who worked for Campbell's Soup Company in New Jersey. For many years the two had a commuter marriage. And then one day, when the professor was visiting his wife in the Garden State, he went out to dinner with an executive in Campbell's R & D division who casually discussed with him a problem they were having in the factory. The professor, whose expertise lay in optical sensors, immediately figured out that there was an intersection between his research and the problems that needed to be solved in the plants where they made soup. Skutlartz proffered a solution to the manufacturing query made by the Executive, and that's all it took: The conversation turned into a series of consulting gigs and project work, and within a year, he was hired full-time by the company.

He was the first PhD physicist (if not the first person with a physics degree at all) who had worked for the food manufacturer. "The vice president who hired me took a fairly large risk," Skutlartz told me. "There were people who were dead set against hiring me. They were used to chemical engineers working as process engineers, but not physicists." Furthermore, as he later found out, a PhD "can be the kiss of death in the food industry," he says, and it is even rarer to find a leader in the industry with a PhD in physics.

But the physicist prevailed at the firm. "I did what the company couldn't do," he admits. "If they needed a crazy idea they would come to me. I wasn't working in the food industry for a long time and therefore could look at the problem from outside the box." And those crazy ideas, which included using optical sensors and x-ray machines to analyze mushrooms for clarity and color and to ensure safety, were all brand-new to Campbell's. As a result of Skutlartz's improvements, he saved the company millions of dollars and went on to win multiple internal awards for innovation.[i]

The moral to this story is:

- If you know your value …
- and can clearly define that value …
- to others in their language …
- so they understand you can solve their problems,
- you *will* gain access to hidden opportunities …
- even in sectors, industries, and fields that appear to be completely disparate from your own …
- and will even be able to make your own opportunities …

just as Skutlartz did. Note: There was no ad that announced "Wanted: Someone to solve this factory problem," or "Physicist needed to provide soup-based solutions." His job didn't even exist. Campbell's created both the consulting and the full-time positions just for this gentleman because, given the opportunity, he was able to appropriately communicate his value to decision-makers who recognized that value and its benefit to them.

To effectively network, and leverage your networks to find the career of your dreams, you have to examine the unique problem-solving capabilities that define your value and then start thinking about how you can use those skills in other industries besides academia. This is not an easy task; it takes extensive self-exploration and external research, informational interviews, time, and even more self-analysis. But know this: The investment you make in discovering your unique value and being able to articulate it to career decision-makers will pay off. You will find not only one career option that gels with your interests, goals, and values, but rather n options, where n is much, much greater than 0. We will go into many of these required undertakings in this book.

Here's how you start: The Skill Inventory Matrix.

Conduct a skill inventory. This is a self-assessment tool that you can utilize to analyze and determine what opportunities are good for you to explore and pursue, and what opportunities are unrealistic given your goals, interests, and skills. Before you begin to complete it, know this: This is a private document. You won't ever need to show this to anyone. It is a personal tool designed to help you articulate your true value which you can use to populate your résumé or CV, cover letters, introductory emails, profiles on LinkedIn and other social media sites, and any other self-marketing document and communiqué, throughout your career. So be truthful, thoughtful, and thorough as you endeavor to fill it out, because it will provide you with extremely powerful information that will arm you to make the right career choices for you and only you.

It is also a living document: The more experiences you have, the more projects you complete, the more jobs or assignments or gigs you pursue, the more information you can glean about yourself. So keep this tool handy throughout your entire career so that as you gain accomplishments you can add them to be able to more methodically and authentically tell your own value story.

A word about cultural issues pertaining to clarifying and ultimately communicating your value to others: The standard in many cultures around the world is not to speak of your success, value or skills. But even in these cultures, if you want a job, you will have to somehow tell the other party what you can do for them. So as you think about your Skill Inventory Matrix, don't hold back on listing the skills you have gained and the achievements that led to that mastery. As we will discuss below, there are appropriate channels and manners in which to promote your skills that take into account that culture's professional ecosystem and its norms for doing so.

Experience	Technical skills	Business skills	Soft skills	Characteristics	Love, love, love	Hate, hate, hate
Job						
Research assistantship						
Teaching assistantship						
Short-term research project						
Volunteer experience						
Leadership experience						
Community-based experience						

Figure 2.1   The Skill Inventory Matrix.

Experiences

In the first column, make a list of all the experiences you have had – jobs, research assistantships, teaching assistantships, volunteer experiences, committee assignments, and even the part-time positions that are far outside your discipline. Think broadly because an "experience" can be almost anything: It can be a job in which you were paid or a volunteer experience, it could be a research-based position like a research assistantship or a teaching appointment. It can also include leadership experiences you have had in clubs and non-profit organizations. And it can most definitely include experiences and jobs that you had that you think don't relate to science or engineering at all. For example, remember that job you had for two weeks at McDonalds when you were 18? You can list it here, because it too will provide you with critical information about yourself and your skills.

Technical skills

Now start thinking about the technical skills you gained from each experience. Technical skills can include:

- "Science skills": The skills of being a scientist, conducting scientific research, or doing scientific problem-solving, including research methodologies, experimental design, data collection, data analysis, and field work.
- Laboratory skills: The skills of utilizing certain equipment in a laboratory or data collection environment; for example, electron microscopy, histology, or in astronomy, certain types of telescopes.

- "Engineering" skills: The skills of solving engineering problems.
- Machining skills: The skills of building, fixing and/or maintaining research and scientific-problem-solving equipment.
- Computing skills: Including platforms, software packages, specialized apps, and languages in which you are adept.

Business skills

Next, think about the business and soft skills you learned in those same experiences. It may seem strange to think that you've acquired, and even mastered, hard business skills from studying science and engineering, but you have, and they are recognized and treasured by employers.

Here's an abbreviated list of the business talents you may already possess:

- Project management.
- Accounting, finance, and budgeting.
- Human resources and training.
- Procurement and inventory control and analysis.
- Risk management and safety assessment (as in ensuring the safety of yourself and your team in the lab or in fieldwork).
- Customer service.
- Sales and marketing (as in convincing your colleagues and potential investors and granting agencies that your work is significant).
- Public and media relations.
- Event planning and management.
- Grant writing.

There is another set of business-related skills which is important to articulate: Your linguistic abilities. If you are bilingual or trilingual, if you can read, write and/or speak another language, you immediately have a competitive advantage in many arenas (after all science and engineering are global enterprises). So think back to your experiences and recall in what languages you can converse and your level of fluency. This will be especially useful for you for opportunities outside of academia: More and more companies are engaging in multinational transactions and if you can solve their scientific or engineering problems *and* dialogue with a Chinese client in their native language, you are providing specific value on several fronts and become the more desired candidate for the job.

Soft skills

You have also gained "soft skills" throughout your many experiences and here is where you list them. According to the Oxford English Dictionaries, soft skills are defined as "personal attributes that enable someone to interact effectively and harmoniously with other people."[ii] I like how Wikipedia characterizes soft skills, as "behavioral competencies. Also known as Interpersonal Skills, or people skills, they include proficiencies such as communication skills, conflict resolution and negotiation, personal effectiveness, creative problem solving, strategic thinking, team building, influencing skills and selling skills, to name a few."[iii] There is no doubt that you picked these skills up while learning science and engineering. I would add to this list diplomacy, management, and leadership.

Characteristics

Characteristics are the self-assessed attributes that you either gained or became aware of their existence from your experiences. These could include your ability to work well against a deadline, or a natural tendency to be detail-oriented or results-driven. You might also find that you function more effectively on independent projects versus those that involve a large team, or that you enjoy complex or nonlinear problem sets versus something more simple. These characteristics may be challenging to identify and qualify but it is an important task to take as you build your Skill Inventory Matrix because this list will give you insight into the type of ecosystem in which you work most productively and satisfyingly. So as you recall and examine your different experiences, be very granulistic about what you experienced and what it demonstrated to you about yourself.

The Love column

This column allows you to really dig deep into your psyche and remember what brought you the most joy while pursuing these different experiences. It will help you concretely see not only where you are talented but also what types of problems you enjoy solving and what environments you enjoy solving them in. Did you love:

- Your boss? Why? Was it his communication style, or his demeanor, or the way he managed the group? Be specific.
- Your colleagues? What was it about working with them that you liked?
- Your group dynamic?
- Your working environment?
- The location of your job – be it the actual lab or facility, the organization, or the city?
- The skills you used? What were they? Why did you enjoy them?
- The instruments/programs/implements/accoutrements of the trade that you utilized to solve your problems?
- The type of problems you had to solve?

- The challenges that you had? What was it about the challenges that brought you pleasure?
- The benefits that you had – including new skills you gained, and even employment benefits.

Very few people undertake this type of exercise in their lives, in which they examine what it is about their career experiences that brought them pleasure. But again, by doing so you are enabling yourself to more effectively pursue or make your own opportunities and to convince decision-makers that you are the right person for the job.

The Hate column

This is the most important column and is your chance to be completely honest with yourself. As I wrote above, as a STEM-educated professional, you have seemingly endless possibilities for career paths, sectors, jobs, and even regions in which you could live and work. So how do you possibly narrow it down? By determining what it was that you hated from previous experiences. This makes sense – after all, the only data we can gather is what we have done in the past. So take a look at all of those experiences and try to remember in vivid detail what it was that you absolutely loathed.

Let your hatred spill out on your worksheet. Did you hate certain tasks, goals or missions? Did you hate certain types of work environments or team compositions? How about the communication and leadership styles of those around you? Or maybe you absolutely hated the location where you worked and would rather stick a fork in your eyeball than have to live there again. These are all extremely useful data! By classifying what you truly abhor, this allows you to put boundaries on the potentially endless career opportunities for you to consider.

For example, let's say for your postdoc you worked in City X. And while you were there for the 2–3 years, you got to know City X and you realize the City X sucks. It sucks big time. You hate it. You loathe it. You couldn't wait to get the heck out. This is *good* news, because now you know something crucial about yourself: You don't have to look for jobs in City X and if an offer were extended to you that involved you living in City X you could immediately dismiss it and go on to the next opportunity. Similarly, let's say that you have also established that you hate a certain industry, in this case for moral reasons. Then if you are given the chance to interview at a company or at a trade organization that represents this sector, you can easily dismiss it because it would go against your moral fiber and you would hate it there.

And a final word about knowing your true hates: The more you are honest with yourself, the better targeted you can be with your networking and the more productive you will be in your career. You can't expect yourself to be successful (i.e. efficient and fruitful) in a company, organization, sector or region that does not align with your values or is representative of an employment ecosystem that you don't like. So even if you are offered a job in that sector and it comes with a hefty salary, while you certainly could consider it, consider this: If you hate what you are doing every day and you hate the values of the organization that is paying you, then you will cease to be productive and you will end up losing your job anyway.

For both the Love and Hate columns you will start to see patterns emerge. Take a pencil and circle Matrix inputs that are repeated over several experiences. Those patterns will serve as serious signposts to assist you in making customized networking, and consequently career, decisions.[iv]

Brand, Attitude, and Reputation: The Networking Trifecta of Triumph

Your brand

> TIP: There is not one public; rather there are many publics who can influence your vocational decisions and paths.

Now that you have completed the Skill Inventory Matrix and can grasp your extensive value, we can now begin to formulate ways to convey this value to the many publics who can influence your vocational decisions and paths. This is the crux of branding.

For years I thought a "brand" was some amorphous advertising term with no real definition. I assumed that it referred to a logo, or a tagline or even a cartoon character that was representing a product or company. But in the early 2000s, I attended a business marketing conference that altered my entire perspective on brands and branding and, as a result, changed the way I pondered business and career decisions.

A brand is not a logo. It is not a tagline or a cartoon character or the name of a product or service or company. A brand is simply a

> TIP: A brand is a promise of value.

promise of value. That's it. When I learned this, I immediately recognized that brands are important to me, not because I am interested in advertising products, but because I am interested in promoting me and my value to potential collaborators. I realized I have a brand, a promise of value, and my biggest mission in life is to ensure that others know and understand that promise as it relates to them and their problems.

Brands are about promises of value, meeting and exceeding expectations, and providing consistent excellence.

You have a brand too, and that brand will serve as your cornerstone for networking and promoting yourself to potential collaborators. But before we go into what your brand consists of, it is important to amplify that brands are about promises, and thus expectations. People interact with brands (products, services, organizations, and professionals) because they expect that there will be a consistency in excellence from them. An easy way to think about this is from the product side. For example, I like Coca-Cola brand soda. I know that if I order a Coke in Tucson, Cairo, New York or Mexico City, on an airplane or in a grocery store, that for the most part, the Coke is going to have the same flavor (allowing for regional differences in water quality, particulates, and bottling practices). My brand loyalty to Coke began at some point in my life, when I first tasted a Coke and enjoyed it. The loyalty continued when I drank Coke after Coke over time, and each time it tasted the same to me and provided a positive experience. Coke delivered on its promise of value and continues to do so in an affirmative way, so I continue to buy it.

To me, the Coke brand represents deliciousness. It is something I enjoy and something I look forward to engaging with. So in my mind, Coke's brand is a positive brand. And where there are positive brands, there are also negative brands, and are perceived so by different people. Just as I love Coke, I hate Pepsi. I have tried Pepsi many times, and each time, I find the flavor revolting. Pepsi has a brand and to me that brand is one of disgust. So given the choice between Coke and Pepsi I will always choose Coke, for the same reason I will always refuse Pepsi – my perceived promise of value from Coke is much better than that of Pepsi.

TIP: Your brand is your promise to deliver excellence, dependability, and expertise in whatever you do.

So Coke is a brand and Pepsi is a brand. But more importantly, you have a brand. That brand is simply your promise to provide excellence, dependability and expertise in whatever you do. You identify what your brand is by utilizing the Skill Inventory Matrix to conduct thorough self-assessment examinations of your unique problem-solving abilities, combined with your skills, experience, expertise, credentials, and even pedigree.

But keep in mind that your brand represents not only your promise of value, but also how exactly you deliver that value. Your unique blend of problem-solving abilities, talents and experience allows you to deliver on your promise to employers, partners, and other types of collaborators with whom you network in a singular way. Your brand is different than others, and your mission is to ensure potential partners know what makes you distinctive and why they will benefit from interacting with you.

Brands are the crucial key in networking and accessing the Hidden Platter of Opportunities. People make decisions about whether to engage with you based on how you present your brand. For that reason, it is

> TIP: Every interaction you have with someone reflects on your brand.

critical that you communicate your brand right from the start. Every interaction you have with someone (and more broadly with your community or industry) reflects on your brand and teaches others what your brand is.

Once you identify what your own promise of value is, you can start "branding" yourself as a leader in your field by performing various self-promotion activities (see Chapter 4). And as you promote yourself and your brand, people will come to realize that there is value in collaborating with you. And that is when you get offered the Hidden Platter of Opportunities.

Your attitude

You will often hear that attitude is everything, and this is very true. I can hire you for a job and teach you all the skills and tools needed to solve the problems of the job. But the one thing I cannot teach you is to adopt and always have a positive, hardworking attitude. Google defines attitude as "a settled way of thinking or feeling about someone or something, typically one that is reflected in a person's behavior." Zig Ziglar, one of the most famous professional speakers and salesman, said, "Your attitude, not your aptitude, will determine your altitude." I could not have said it better myself.

Your attitude is the calling card of your brand: It verbally and non-verbally communicates your brand to others. It teaches people what you stand for and it educates others how to treat you and how to perceive your value. And since every interaction you have with someone reflects on your brand, your attitude must always be positive. If it is not, others will perceive your brand as something it is not – perhaps something that is negative and not of value to them.

This concept can be sometimes challenging to understand, particularly in the sciences and engineering. After all, we are taught as early as our undergraduate studies that what is most important to progress in a STEM career is your STEM outputs, as mentioned above. And we see, comically in academia, professors with "bad attitudes" all the time and yet these people have attained professional success. Although they may have bad attitudes now, I can assure you that when they interviewed for their jobs, they did not display any sort of negativity in their discussions. Furthermore, as they advanced, they may demonstrate inappropriate etiquette or behavior in certain situations, but I can almost guarantee that they did not get tenure based solely on their outputs. In fact their outputs were shaped and propelled by their overall positive attitude towards science or engineering.

But the fact is that people around you at all times are making snap decisions about your brand and the benefits (or non-benefits) that you can provide them based purely on your attitude. You can especially see this manifest itself during networking receptions. You and I engage in a conversation, having never met before. As we shake hands and offer introductions and then proceed to chat, I am consciously and subconsciously making mental notes about the way you are interacting with me. In other words, I am sensing your attitude and then digesting this information to provide me with a conclusion about what your brand is. Your attitude tells me volumes about who you really are, or, more importantly, let's me *perceive* who you are.

> TIP: Perception equals truth in the minds of the publics, and everyone is a member of the publics, including your twin and your clone.

When you walk into a networking reception or a job interview or an informational interview, is your head held high? Are you enthusiastic about being there, do you look the person in the eyes and smile, do you engage them in a mutually beneficial conversation? Or do you look at your feet, chew gum, speak softly, or worse, look away from the person and act like you are bored to be there? Do you employ professional etiquette, and honor and respect the person with whom you are conversing? Like it or not, people notice these things, both consciously and subconsciously. They see how you treat them and they make decisions about you and your brand based on your attitude. And since perception equals truth in the minds of the public (and everyone is a member of the public including your twin and your clone), they assume what they witness is the truth. Your networking (and career advancement) goal is to ensure that what others perceive about you is the truth.

So it is critical for you to always have a positive, professional attitude when speaking with and interacting with members of the publics. We want to ensure that when someone observes your attitude that they are left impressed – by you, your talents, your expertise, your credentials, and so on, which will encourage them to want to engage you again. This is the start of the networking partnership; this is the start of the other party beginning to think about offering you access to that Hidden Platter of Opportunities. But if someone perceives you in a negative light at a mixer, conference or other event, and they know nothing else about you, having never met you before, your negative attitude can immediately cause any opportunities (hidden and advertised) to disappear. In other words, when you honor me with a positive attitude, it is so powerful that I can see myself building a relationship with you and my mind starts to wander to look for opportunities for collaboration. Conversely, if you display a negative attitude my mind immediately closes down, and my singular interest is to get away from you. As the Seinfeld character the Soup Nazi might say: No opportunities for you!

But of course it is not only at cocktail receptions that your attitude convinces people whether they want to partner with you. Attitude goes a long way in converting job interviews into actual job offers. This is even the case in academia, which plays host to a diversity of personalities and attitudes. When you go for an interview at an academic institution, it is important to have a positive attitude and to demonstrate in your words, actions, and even your job talk that you are collegial and are someone that will add value to the department through not only your skills and experience but via your collaborations and discussions as well. When you interview for the job and have conversations with departmental faculty, the chair, and the dean, all of these other parties are trying to get a sense from you about your willingness to contribute to the team. Your attitude is the marker that provides them with this strategic information. Of course, this attitude towards attitudes extends beyond academia to all other sectors.

So in both job interviews and informational interviews, you must sustain a positive, collaborative, and professional attitude throughout the experience. You do not want people making negative decisions about your future based on any negativity they may perceive. And people do this all the time. I know I have. I recall speaking with an astronomy professor at a university mixer one afternoon. We were having a pleasant enough chat although I did notice his eyes were not focused on me as we spoke and instead were watching other parts of the party. Suddenly there was a commotion over to the side and, naturally, everyone instinctively turned to look. When I turned back from the big bang a few half seconds later, the astronomer was gone. He had used the hubbub as a means to skedaddle. This was a professor I had known for almost 20 years, so I recognized that although his behavior was unprofessional (he should have at least said goodbye), he was probably just uncomfortable in the situation and antsy about getting back to his lab. But what if I was his current boss or potential boss? I could have taken offense at that poor display of attitude and made a mental note in the future to not invite him to other events where he could possibly embarrass me or himself, thereby effectively shutting the door on any possible hidden opportunities that I could provide him down the road.

> **TIP:** Being professional simply means being serious about your craft at all times, in all ways.

The following is another example of someone (in this case me) taking notice of another's attitude and therefore appraising their brand based only on information attained from their behavior. My first real job after graduating from college was as Director of Communications for the University of Arizona (UA) Department of Physics. I was in charge of all public and media relations for the division and I needed to hire a student assistant who would help me with various communications tasks. In essence this person would serve as a face of the department (and by extension, the department's brand), so I was especially eager to hire someone who was very professional.

So I did something in the interview process that many hiring managers do, both consciously and subconsciously. I introduced a "stimulus" into the interview scenario to see how the candidates would handle themselves in the situation. I certainly didn't invent this idea; I had heard of a multinational semiconductor company taking their candidates for sales positions out for lunch as part of the interview and pre-arranging with the restaurant to mess up the meal in every way possible. The interviewers were watching how the candidate would respond to the stimuli, such as dirty utensils, rude wait staff, and an incorrect food order delivered to the table. They were waiting to see how his attitude manifested itself when faced with these inconsistencies, and they were using this information to decide whether this candidate could be a valuable asset to their sales team.

In my case, when candidates arrived for their interview with me, I simply ensured that the door to my office was closed. I wanted to see what they would do: Would they do the most professional thing, which was to knock on my door, wait for me to say enter, and then do so? Would they run away or crawl in a ball on the floor crying hysterically? Or would they walk right in?

I was satisfied that almost every single person that day took the most professional action and knocked. But there was one kid who did not do this. And to this day, more than 16 years later, I remember him more than everyone else, including the person I hired (in fact, I don't even recall whether I hired a male or female). The reason this student is imprinted on my brain is entirely because of his negative attitude. When he arrived for his appointed interview time, instead of knocking on my door, he walked right in. And then he sat down in the chair across from me, which was one of those office chairs that rock back and forth. And as I began the interview, he put his hands behind his head and rocked back and forth in that chair with a smirk on his face. He clearly thought he was cool. He clearly thought he was awesome. But what he clearly didn't know was that his attitude was poisoning my opinion of him with every nauseating second. In fact, his attitude clearly communicated to me that he was a cocky moron, untrustworthy and could potentially be an embarrassment for me and the department. Based on his attitude, my decision was clear – I didn't hire him.

> TIP: Attitude is everything: It instantly acts to open or close access to career-changing opportunities.

But of course, I recognize that attitude is only one part of the package. In fact, he could have been the most brilliant public relations strategist and a superior problem-solver who could have enhanced my office tremendously. But his attitude communicated something different. His attitude told me a story about his brand that may have been false. And without knowing anything else about him, I made my decision based on the only piece of data that I had – his attitude.

So to be clear, attitude is truly everything. And when you are commencing your networking strategy, with a goal of crafting these mutually-beneficial collaborations over time, it is imperative that you use your attitude to convey your promise of value to the other party. If you don't, the exact opposite will occur – your attitude will demonstrate a perceived promise of negativity and possibly even a deficit if they were to partner with you. Doors will close; the Hidden Platter of Opportunities will disappear altogether.

Reputation

The final piece of The Networking Trifecta of Triumph is your reputation. In a nutshell:

Your reputation is your most important asset.

This is almost contrary to what we are taught in academia, where for the most part people attest that it's what you know that will get you the job. But truth be told, it is not what you know, it is what people know about you and your knowledge, abilities, brand, and attitude that will get you the job and other game-changing career opportunities.

Your reputation is what people know about your brand and your attitude. It is an extremely important piece of networking because it is the carrier across time, space, and extra dimensions of what value you can provide other parties, teams, and institutions.

> **TIP:** Decision-makers want to hire "known quantities," because it reduces the risk associated with bringing in talent.

The reason surrounding this makes sense; people want to engage or hire "known quantities" – people whose brands and attitudes are already known in the community. Your reputation is how others confirm that you are a known quantity. The more you are seen and the more positive bits of information people know about your brand and attitude, the lower the risk they have in engaging or hiring you. Given a choice, I would rather hire someone whom I either know personally or by reputation, or who others in my networks know and can vouch for by either personal knowledge or reputation. I would prefer not to engage someone who is a stranger to me and whose reputation is either unknown to others I trust or is negative in some form.

About 10 years ago, I was attending the monthly meeting of the Southern Arizona chapter of the Public Relations Society of America. This was a regular event for me – to meet new people, to exchange information, to learn new ways of solving PR problems I might find in my job. Each meeting incorporated an invited speaker as well as time before and after the meeting to freely chat and network with those in attendance. At this particular session, the presentation consisted of a panel of editors from local publications offering tips on how to pitch them stories. One of the speakers was in a new job – he had just taken over as the editor of the regional business newspaper, to which I subscribed and read regularly.

> **TIP:** When there's new blood in an organization, this is often the best time to pitch new, or even create, opportunities.

When the formal part of the meeting concluded, I practically jumped from my chair and pushed my way across the room to introduce myself to him. I thanked him for his speech, congratulated him on his new position, and offered him my business card. I then blurted out that I had an idea for a column that I would like to write for him and inquired whether he would be interested in meeting with me to discuss it. Perhaps because he was so new on the job, and perhaps because he himself wanted to get to know other professionals in his sector, he granted me the appointment.

I approached the meeting as if I was going for a job interview – which of course, it really was. I took with me a portfolio that included an example of a potential column, previous writing samples, my résumé, and other projects that demonstrated my proficiency in public relations and my ability to solve his problems (relating to upping his readership and ad revenue) with finesse. What I specifically did not include was a list of references.

We had a very pleasant conversation and, of course, after I left I sent him a follow up email to thank him for meeting with me and to tell him I was looking forward to working with him. And then I waited. I waited in fact, several months. And then one day, I received an email from the gentleman stating that he wanted to move forward on the column idea, and that I have "an excellent reputation."

Of course I was thrilled. Of course I was excited. But I was also completely perplexed. How did this guy, who although not new to the city was new to his job, know that I had an "excellent reputation"? How did he find out and then conclude that I would be an asset to his publication? The answer was obvious once I thought about it: He clearly spoke with people in his own networks, professionals he trusted, and they must have assured him that my brand and attitude were something that would bring him benefits. Our networks clearly intersected and my reputation was carried from one to the other to him. And since it represented me as someone who could be an advantage to his company, he made the decision to hire me.

That's just one example of how powerful your reputation is to crystallize access to hidden opportunities and to actually secure you those opportunities. And since you now know that your reputation is your most important asset, keep these thoughts in mind:

- Your reputation can open or close doors instantly. 'Nuff said.
- It's a small world, after all. Your industry is tiny, your field is tinier, your subfield is practically picometers wide. One of the reasons

networking is so powerful is because we are so closely connected to everyone else, more so than you might realize with a cursory glance. And of course the theory of six degrees of separation proves this once and for all.

- People talk. It's human nature to speak about others, even if we are taught that is not necessarily good form. Regardless, people talk to and about each other, especially when they are looking to expand a collaboration, bring in new talent, or find new inspiration. They ask for recommendations and they encourage others to opine about on whether they should be in business with you. Make sure that when "they" talk about you, they are only saying good things. As a corollary to this, we don't want people saying negative things or, possibly worse, nothing about you at all because you are unknown to a community.

- People (often) believe what they hear. When a trusted member of a community tells another that you are someone who will add value to their group, they believe it. Similarly if someone says something negative about you, the other party will believe that too.

- Your reputation will pop up in places you'd never expect. A reputation is a funny thing – it's neither solid nor liquid nor gas nor plasma, but somehow it slithers and sneaks into nooks and crannies in the "ether": It reaches, and more importantly influences, people, institutions, communities, even regions that you could never have planned for. It's out there, it can have a life of its own and you really cannot predict who is a "carrier."

- Know what is being said about you at all times. You have to keep your reputation clean, pristine, and truly reflective of your brand and attitude at all times. This requires constant research and maintenance. Google yourself. Talk to your trusted mentors and friends regularly. Find out what is being said about you.

- You must engage in constant reputation management activities. Your house will become dirty and start attracting ants if you don't regularly clean and maintain it. Similarly, your reputation runs the risk of becoming either sullied or fading out completely if you don't maintain it. The publics must consistently perceive you as a contributing member of the community and this requires you to consistently engage in self-promoting activities and to keep networking.

- If there's a problem, try to right it, right way. If you determine that something is askew with your reputation or that your reputation has taken a negative hit, you must tackle this problem immediately. Endeavor to find out what the source of the negativity is and see if you can counteract it with professional outputs and other activities. Don't wait for or expect others to right a reputation wrong – you have to take the initiative to find out what the problem is and determine avenues to bring it back on course.

I gave a speech a few years ago and discussed these concepts and how we have to ensure that our reputation is absolutely sterling at any point in time. A woman raised her hand to ask my opinion about a very unusual but deeply disconcerting image problem: She has the same name as a serial killer. So when people Google her, they might assume that she is a murderer. This is no laughing matter to this unfortunate student – I shared her concern and sympathized with her, as this was not the first time I had heard of others being mistaken for another who has a bad reputation of some sort.

My proposed solution for the woman was to start using either her middle name or her maiden name (if she was or had been married) as part of her moniker in all professional documents and correspondence. This included her CV, published papers, presentations, LinkedIn profile, institutional and personal website and directory, Twitter and Facebook accounts, and so on. I also recommended that she change the name on her email account to reflect the new name, as well as the signature on all email. She could have even gone a few steps further and created a blog under her new name, and then injected it with tags containing her full name. The goal was to create multiple "imprints" with the publics, so that soon her STEM outputs and networking correspondence were directly and solely linked with "Jane Oppenheimer Doe" and not the murderous "Jane Doe."

Your Brand Statement/Elevator Pitch/30 Second (or Less) Commercial

We have established the breadth and depth of your talent and what your specific blend of skills and expertise is. We now have to prepare to announce it and your presence to the publics. This is the purpose of a brand statement. Also known as an elevator pitch, this short and quick self-promotion "commercial" is designed to:

- communicate what you are talented in,
- articulate what benefit you provide,
- convince people why they should partner with you – for a job, internship, fellowship, grant, or other opportunity (now and in the future),
- entice someone to learn more about you,
- provide the other party with information that inspires them to visualize hidden opportunities in which you could collaborate.

It's called an elevator pitch because it evokes the following scenario: Imagine you get on an elevator with Bill Gates and you have until Floor 16 or so to speak with him and convince him that you could be a vital ally for him. How do you do it? You deliver your brand statement, or elevator pitch. It's a quick synopsis of what you can do and why and how you can do it for him.

The brand statement should include:

- your unique assortment of skills, experience, expertise,
- your problem-solving abilities,
- your overarching goals,
- the benefit you provide the other party → your value,
- your competitive advantage.

You will deliver your brand statement many, many times throughout your career to myriad publics, some of whom are in your subfield and field, some who are not, and some who are completely outside the realm of science and engineering. The more you deliver it, the easier it will get to do so and to adapt it for different audiences.

Your Skill Inventory Matrix will be a great asset for crafting your brand statement. All of those skills and interests that you determined you possess from that tool can now be used as a scaffolding to design your brand statement(s).

What do I include in my brand statement?[v]

- Introduce yourself: State your name slowly and clearly so that the other party can hear you and understand it. This is especially important if your name, like mine, may be difficult to pronounce, or hard to hear if you say it quickly. Many times I have introduced myself as "Alainalevine" and people don't catch the specific pronunciation or even that I have a last name. So give yourself a nanosecond or so between your first and last names so the other party knows who you are.
- Describe your area of expertise: for example, "I am a wildlife ecologist with an expertise in megafish that are found in fresh water bodies around the world" or "My background is particle physics and I am currently doing a postdoc at CERN."
- State a strength or skill in which they would be interested. This may be a challenge if you don't know who your audience is yet for your brand statement. But over the course of your conversation, clues will be revealed which will give you hints as to what skills that party would like to know about. However, if you are delivering your brand statement at a conference related to your subfield, you will already know what skills the other parties will desire to know, so you can prepare your brand statement based on what skills the subfield finds relevant, useful, and valuable.
- Follow that with an accomplishment (or two) that proves you have that skill. Don't be afraid to state your credentials or something that truly

separates you from the rest of the pack. "I recently completed a fellowship in the White House Office of Science and Technology Policy," or "I was part of the team that found and authored the paper on the Higgs-Boson particle."

- Discuss a goal that the other party might be interested in: Think about what you are hoping to communicate and the hidden opportunities you are aiming to unlock. "I am graduating with my chemical engineering master's degree and am interested in pursuing opportunities in the petrochemical industry" or "I write about physics for magazines like *Scientific American* but am thinking of transitioning to science outreach."

The punch line: the benefit you can provide

Most importantly, your brand statement should convey a benefit that you can provide. Remember, networking is about not only accessing hidden opportunities for yourself but also finding hidden opportunities for the other party and providing them value so they can solve their problems. So this is truly the moment when you get to shine the most and demonstrate your true motivation: To help the other party solve their problems. Think of this aspect of your brand statement as an avenue to elicit a response from the other party. Some of the phrases that you can incorporate to achieve this can include:

TIP: People truly appreciate when you are resourceful in recommending others for opportunities that you yourself cannot fill. It can go a long way in solidifying the networking partnership.

- I can be/could see how I can be of immediate benefit to your company because …
- I can see that there is synergy in our areas of research in that …
- I could assist you with your efforts because I …
- I would like to *explore* the potential to partner …
- I can see there may be an opportunity to collaborate because/in/in this way …
- Perhaps we can *explore* working together/doing X together …

Some notes about design and delivery of the brand statement:

- It will be customized depending on the audience, mode of delivery, and scenario: You will never have one brand statement. You will actually have many, many brand statements that will be communicated in many different ways and customized for each audience. For example, although I have expertise in science writing, when I was speaking with the wife of the senator on the airplane, I only discussed my speaking and comedy background. From our conversation, it occurred to me that this would be the relevant information to share, as opposed to other areas of my brand. Similarly, when I attend the annual conference of the National Association of Science Writers and meet editors at mixers, I talk mostly about science writing because that data is what is most important for them to know. See below for more details.

- It may not happen all in one shot: Very rarely do you have the opportunity to actually deliver an elevator pitch in 20 seconds or so. (The most obvious exception to this rule is at a career fair, when you do state your full elevator pitch to the recruiter.) Most often it is gradually revealed as you converse with someone. It involves a give-and-take approach, in which you customize your brand statement for the person with whom you are speaking, based on their own answers to common questions like "what do you do?" and "what organization are you with?"

- You can reveal your value in the form of a question: "Oh you work for Microsoft? I am completing my master's degree in electrical engineering with an emphasis on embedded systems. What kinds of system architecture do you use for embedded systems? I like to use X."

- You should ask "exploratory" questions: This too is tailored based on what the person reveals in the dialogue. For example, after they disclose that they are a science journalist with X magazine, you can respond with "I really enjoy writing about (my field) and how its innovations have sparked applications in everyday life. Have you thought about doing a series of articles about this type of thing? Perhaps I could assist you in this effort."

- You should demonstrate your resourcefulness: YOU don't have to always be the answer to the question. Instead, if they divulge that they need someone who has a mechanical engineering background, and you do not possess this background, you can share that you have contacts with the American Society for Mechanical Engineering and would be happy to make some introductions over email for them.

Start by having a few different versions:

- One that is highly technical – for someone in your subfield.
- One that is technical – for someone in your field but outside of your subfield.
- One that is moderately technical – for a scientist or engineer who is not in your field (for example, if you are a biologist and you were speaking to an electrical engineer).
- One that is even less technical – for a potentially science-educated public (for example, the readership of *Scientific American* consists of

people who are enthusiastic about science and maybe have some science or engineering education).

- One that is strictly for a lay public – interestingly, this will be your most valuable brand statement and quite possibly the one you deliver the most. When you communicate it, you may even end up starting with pieces from this version and then switching over to another version as the audience's knowledge, expertise, and fascination reveals itself. The best feature of crafting a lay person's version of your brand statement is that it gives you priceless practice in communicating your high-level value to those who don't necessarily understand the details but could still form a mutually-beneficial partnership with you. You will have to deliver this version in this manner many times throughout your career – for example, to recruiters or human resource professionals who do not understand the scientific or engineering significance of your experience, but still have to make a decision about you as to whether they should present you as a candidate for a technical job. You have to convince them that you are the right person, who is smart and technically talented, but you can't necessarily use jargon, acronyms, or certain phrases that only a professional in that subfield would know. If you can persuade the HR rep that you would be able to solve the company's problems using a brand statement that gives her strategic information and also honors her (don't "dumb it down" – just clarify in her language what you can do for her), you have made her job easy and will be able to move forward in the interviewing process.

Scenarios in which you will deliver your brand statement:

- At a networking mixer at a conference or in your community.
- When introducing yourself to a speaker at a conference.
- If you are giving a poster presentation at a conference.
- At the exhibition hall of a conference.
- At a career fair.
- In an informational interview.
- In a job interview.
- To a new colleague.
- To the department head, dean, or president of your institution.
- To the colloquium speaker
- At the Journal Club.
- To potential investors or funders of your work.
- In outreach programs – to K-12, teachers, parents, and other potential lay people.
- At a gathering of hobbyists.
- On an airplane, in line at the grocery store, at a bar, and so on.

Ecosystems in which you would deliver your brand statement:

- In person.
- On the phone.
- Via Skype/video chat.
- In an email.
- In an online job application.
- In a cover letter for a job, fellowship, grant, or award.
- Via social media: Your LinkedIn profile, Facebook page, and Twitter handle.
- In a blog or your portfolio webpage.

Here are some thoughts to keep in mind when building your brand statement.

Practice, practice, practice: The more you practice, the better you will get at delivery and also adaptation for new audiences. And the best place to practice is in low-stake networking situations like on an airplane or a grocery store where you also have the added task of ensuring that the party, who is not a subject expert, truly understands and recognizes your value. One of my favorite places to practice my brand statement is at a career fair. The career fair is great for networking. Most universities and many conferences hold career fairs. And even if the fair is not devoted to STEM fields, it is still a fabulous opportunity for you to get used to delivering your brand statement in such a way that causes others to take action. It encourages you to become adaptable and nimble in your delivery since each company is looking for a different set of skills. Furthermore, the career fair ecosystem fosters a system of a conversation, so it also gives you a chance to follow up and expand upon your brand as you listen to the other party talk about their needs for an employee. I have attended many, many career fairs and have found them to be extremely helpful. And here's a tip – just like with networking at an event, start with a low-stake booth to bolster your confidence. If you have no intention of ever working for an insurance company, but want to work for Intel, start your career fair experience at the insurance booth. This is exactly what I did when I was getting ready to graduate from college. By the time I got to my goal booth, which was in fact Intel, I had given my brand statement five times to five different companies for which I had no inclination for working. My Intel delivery was smooth, to the point, clear, concise, and it automatically welcomed the other party to ask more questions, which made the engagement more enriching for both of us.

It should feel as it comes naturally: for both you and the audience, it shouldn't sound or feel like you are delivering a prepared speech that

> **TIP:** Career fairs are excellent for networking, both with potential employers and other job-seekers.

requires cue cards. It should be a natural expression of your value and your passion for your enterprise. People really appreciate others' excitement and enthusiasm for a subject and they take note of it. I was recently at an event where astronomers were giving lectures for a lay audience made up of donors for a certain observatory. The last speaker was a young woman who was so engaging and happy to be up there discussing her scientific pursuits and discoveries, and she did it in such a way that the audience truly understood the relevance to her and to them. As she concluded and asked if there were any questions, someone raised their hand and stated "I can see how much passion you have for this subject. Thank you for sharing it with us!" Even though she was not delivering a brand statement per se, her entire presentation was in essence her brand statement, and she articulated it so naturally and with such enthusiasm that the audience couldn't help but become more intrigued about her research. She probably even helped the observatory raise more funds that day!

> TIP: Sometimes, your brand may not be delivered in a single "statement" – it could be delivered over the course of a conversation or in a presentation.

Speak up: If the other party can't hear you, whether it is because it is a noisy room or because you speak softly, you are losing a potentially precious opportunity. This doesn't mean you should shout it out, but make sure your voice is heard. And there's another advantage to literally being heard: It denotes confidence, poise, and professionalism. The more you practice this, the more natural it will feel and the better you will become at delivering your brand statement in such a way that the words and the way you speak show the other party you are ready to assist them in solving their problems.

Be prepared for follow up: Any comment that you make as part of your brand statement is fair game for someone to ask follow up questions. In fact, if you craft it correctly, you'll find that it logically elicits questions, which is a great thing to have in networking – it means that you have enticed the other party to learn more. So be nimble and flexible in your delivery to allow for people to interrupt you. Go with the flow.

Stand by your brand: Don't say anything in your brand statement that you can't completely back up – in other words, don't embellish the truth, lie, or mention something that you knew how to do years ago but have since forgotten. I found this out when I wrote my very first CV as I was graduating college. I didn't know what was supposed to be on a CV, so I listed every club that I had attended at least one meeting of. When I won a math award, my CV found its way into the hands of the chair of the math department who invited me to his office for a chat. When we met for the first time, his initial words were "how many hours do you have?" I was clueless as to what he was referring, so I responded "Do you mean credits? I have probably about 130 or so credits." And he shot back "No! How many *hours* do you have?" He was alluding to what I considered to be a very minor entry on my CV, a listing about how I had been a member of the scuba diving club at the American University in Cairo. And in fact not only was I not a member (I had only gone to a handful of meetings), but I wasn't even a scuba diver. I tried it at the deep end of the pool and quickly realized it was not for me. But by listing it there I had opened the door for him to discuss it with me. Needless to say I was embarrassed, but fortunately I learned the lesson. I want to ensure that you always feel confident in communicating your value, so you can periodically do a "brand assessment" to determine if you still have mastery of the skills you mention in your statement. For example, after my semester abroad in Egypt, I took many classes in Arabic and became fairly fluent. But that was more than 10 years ago. So although at one point I did mention I was fluent in Arabic in my brand statement and on my résumé, I don't any more. You always want to make sure when you communicate your brand that it is completely factual and that you have evidence to support it.

> TIP: Periodically, do a brand assessment and ensure that your brand statement (and all your marketing documents) is up-to-date and truthful and not deceptively or accidentally communicating that you have a higher level of expertise than you currently have.

Tailor it for different audiences: As you network and get to know people from different communities, educational backgrounds, and career aspirations, you'll need to adapt your brand statement for each person. This will get easier the more you do it. You will also learn how to use different vocabulary or buzz words to make your point. For example, if I was discussing my expertise in marketing to a client in academia, I would probably reference my ability to design programs that help recruit students. But in the business world I would refer to the students as customers. By using their language, they understand my brand better and more quickly, how I can help them, and that I understand their world and culture. This in turn communicates to them that I can jump in and build a mutually-beneficial partnership that will help them solve their problems immediately.

Keep it simple, short: The bottom line is a brand statement is not a thesis. It is not meant to take three hours to deliver. It is designed to very quickly and as simply as possible communicate what you are excellent in and how that might benefit the other party. It is designed to start a conversation and to serve as the foundation for building a partnership. This doesn't mean you have to dumb it down for anyone – your brand statement can be shared in such a way that it is easy to understand *and* honors the other person's intelligence and level of expertise. But keep it simple, short and to the point and you'll get the best results.

A Brand Statement Case Study: Jerzy Rozenblit, Professor

For an example of someone who knows how to articulate their value and problem-solving skills effectively in multiple brand statements and ecosystems, I couldn't help but think of my friend and colleague, Dr. Jerzy Rozenblit. Jerzy is the Raymond J. Oglethorpe Professor of Electrical and Computer Engineering at the UA. He is also a professor of surgery in the UA College of Medicine, and the inventor of a tool that teaches surgeons how to conduct robotic-aided operations. Jerzy was the chair of the department for eight years and, as a scholar and academic, he constantly has to deliver his brand statement to different people. Here are some ways that he shares his value:

For a potential research partner in computer engineering:

Hi, I am Jerzy Rozenblit. My new, hapitic and augmented reality guidance system for minimally invasive surgery will complement your work in medical simulation very well. We should discuss potential teaming arrangements.

For a potential partner in engineering:

My expertise is in modeling and design of complex, engineering systems with a strong emphasis on software and hardware co-design. The application focus is on medical technologies.

For a potential partner in academia but not necessarily with a scientific background:

I teach and research methods and techniques that integrate high technology in medicine, especially surgery. I focus on how to make surgical training more effective using computer-based assistive tools.

For a potential investor:

I design surgical training simulators and surgical guidance systems. They will revolutionize the way we train students and residents, and the way robotic surgery will be done in the next decade.

For an executive within academia:

I have a demonstrated academic leadership experience and acumen. I have managed a large, multi-disciplinary engineering department for close to a decade, and brought it prominent recognition. I also manage many large grants that develop solutions to cross-cutting research and education problems.

For someone not in academia:

I design and engineer systems for minimally invasive surgical training and computer-assisted surgery. I also teach computer engineering undergraduate and graduate students.

For someone on an airplane:

Hi, I build surgical systems that make operations such as gallbladder or appendix removal safer for the patients. They help you recover faster with much less pain.

As you can see, Jerzy has multiple versions of his brand statement which he has thoughtfully crafted in advance so that when a particular audience presents itself, he can move fluidly from one brand statement to another. He can even combine them based on the interests and backgrounds of those with whom he is conversing. This is the kind of brand statement repository you should endeavor to create for yourself.

Chapter Takeaways

- A brand is a promise of value and everyone has their own brand.
- Your brand constitutes your promise to provide excellence, dependability, and expertise in whatever you do.
- Part of the challenge of networking is being able to identify and clearly articulate your brand to numerous publics so they understand what benefit you could provide a partnership.
- You can identify your unique brand by conducting a skill inventory and self-analysis of your distinct problem-solving abilities.
- Your attitude is the calling card of your brand and people make decisions about your brand based on your attitude.
- Perception equals truth in the minds of the publics.
- Your reputation is your greatest asset: What people know about you, rather than what you know, is what gets you access to the Hidden Platter of Opportunities.
- Your reputation will sneak into networks that you may not have even realized existed.
- You communicate your brands through brand statements that are delivered during networking and through reputation management activities.
- You should have multiple versions of your brand statement, which can be delivered to multiple audiences in multiple ways.
- Practice your brand statements to bolster your confidence and improve your delivery.
- Appropriately and clearly articulating your brand gets you access to the Hidden Platter of Opportunities.

Notes

i. Mmm, Mmm, Physics: The Man with the Plan for Cans, APS News, December 2011, http://www.aps.org/publications/apsnews/201112/profiles.cfm.

ii. http://oxforddictionaries.com/us/definition/american_english/soft-skills.

iii. http://en.wikipedia.org/wiki/Soft_skill.

iv. From Physical Scientists Can Do Anything: Here's how you start your career planning, *Physics Today*, March 21, 2013, http://www.physicstoday.org/daily_edition/singularities/physical_scientists_can_do_anything_here_s_how_you_start_your_career_planning.

v. Adapted from Drexel University Career Development Center website.

3 Determining the Right Opportunities for Me

I hope by now you realize that as a science or engineering educated professional, you have potentially endless career opportunities. So now you have to start putting some boundaries on these opportunities to find the right ones for you. One of the tremendous benefits of networking, and the resulting strong networks you craft, is that it allows you a greater chance to seek and find opportunities that are the best fit for you professionally, personally, and intellectually. Networking can serve as a strategic tool in your career planning.

The Concept and Pursuit of Bliss

Below I discuss just some of the career paths that you can pursue as a science or engineering educated professional, all of which I discovered from networking myself. Of course this is only a partial list. There are many, many jobs you can acquire and even more game-changing career opportunities of which you will become aware and can pursue through networking. But no matter what job or career you desire to chase, you should have one central objective: To find a vocation and a professional ecosystem that will provide you with consistent and sustainable bliss, enjoyment, and intellectual stimulation.

This concept of bliss has been studied and pontificated on for practically eons by scholars and philosophers. But I found that the best interpretation and resulting explanation of what this "bliss" means to you in terms of career planning is simple: You are intellectually and creatively involved in an activity that provides you with its own inherent reward, and is so enjoyable and you are so focused on it that you may lose track of time. You may not even notice negative stimuli or elements associated with the activity because you are using your high level of skills to solve complicated problems; the problem-solving itself is a pleasurable endeavor.

These components of bliss manifest themselves in various ways as you engage in activities that you enjoy. For example, I absolutely love

Networking for Nerds: Find, Access and Land Hidden Game-Changing Career Opportunities Everywhere, First Edition. Alaina G. Levine.
© 2015 John Wiley & Sons, Inc. Published 2015 by John Wiley & Sons, Inc.

public speaking. When I am giving a workshop or a keynote speech, I am in total bliss: I am completely in the moment and fully concentrating on my performance. And as much as I enjoy the audience's positive reactions, I am most happy because I am doing it, not because I am able to make people laugh with my jokes. For that reason, I don't care if I am giving a talk for 2 or 200 people. I get so involved in my speeches that I often lose track of time. At one conference, I had a two-hour window to give a workshop and was having so much fun that I didn't realize three hours had passed (fortunately the audience in this case enjoyed it as much as I – nobody left even though we were one hour beyond my allotted time!). I also have experienced another element of bliss when I have given speeches – I don't notice anything other than the actual speech and my interaction with the participants. I first recognized this in myself when I was booked to give a four-hour workshop. I wore high heels for the entire length of time I was speaking, and had no knowledge that I was in agony from the uncomfortable shoes for the majority of the time I was on stage. It was only when I concluded my performance and thanked everyone that I suddenly felt a shooting pain in my feet and I realized that it had been there the whole time – I just never noticed because I was in bliss.

This feeling of bliss, or whatever you want to call it, is what I hope you achieve in your career. And it is completely possible and probable that you will realize it, especially as you conduct self-reflection exercises like the Skill Inventory Matrix and the SWOT Analysis (see below). And the more people with whom you network, the more opportunities you will discover that seem almost uniquely designed to provide you with professional bliss.

My point is this: You *should* experience bliss in your professional life. And it is absolutely achievable. Your professional goal is to find and distinguish opportunities that will allow you to experience bliss and utter happiness from simply doing your job. Networking will unlock these opportunities for you. I once started a conversation with a woman sitting next to me on an airplane flying from Chicago to DC. As it turns out, we had a lot in common: We were both alumna of the UA, and we both came from the same small town in New Jersey – small world indeed! As I engaged her, I soon learned of her seemingly unusual job, one that I heretofore did not even know existed: She worked in marketing for a major cruise ship line. She was tasked with organizing huge corporate marketing events that showcased the cruise company and their partners. For example, as she described to me, she designed and orchestrated an extravagant event for an automobile manufacturer that was premiering a new car model to the media via a party on one of the cruise line's vessels. This woman was charged with overseeing every aspect of that affair, from the media relations and food and beverage to the security and negotiating the maritime issues and legal concerns, to the logistics of getting

> **TIP:** Your main professional goal should be to find and distinguish opportunities that will allow you to experience bliss and utter happiness from simply doing your job.

the cars and people on and off board. At the time I was early in my career as a public relations professional and was hungry for any information about the different types of careers I could pursue in this realm. By having this enchanting and eye-opening conversation with this person, I discovered an entirely new world in which I could be a communications practitioner, as well as a new world of PR problems to solve and exciting challenges to face in the cross-over industries of cruise lines, automobile manufacturing, and maritime logistics. I couldn't believe the gold I had struck. I never ever would have even thought that this type of opportunity existed; my mind didn't even have a means to imagine this universe. But thanks to low impact networking on the airplane with this gal, I was now armed with strategic information that could take my career to

> **TIP:** You are at your most productive (and in concert, most valuable to your employer and/or partners) when you are in bliss, or are experiencing bliss in your profession.

new and different waters, and enable me to chart a course for professional bliss in an entirely new industry. The lesson is clear: The more you network with people from diverse backgrounds and industries, the more you learn about opportunities where you can apply your skills and abilities in novel ways to find your own personal bliss.

Your Personal SWOT Analysis

So how do you begin to characterize opportunities that might lead you to bliss? One way is to complete a Strengths, Weaknesses, Opportunities, and Threats (SWOT) Analysis. A SWOT Analysis is a common tool you see in business schools, particularly when it comes to examining market potential for a product or service. But it is also an extremely useful tool for analyzing career potential. The point of this exercise is to start determining where you would best fit for a career and as such what networks you should start trying to infiltrate so you can embark on this career.

Notice that the top half of the SWOT are characteristics that are intrinsic to you: strengths and weaknesses that *you* possess, whereas the bottom half are extrinsic: opportunities that you can pursue and threats to those opportunities.

The process to complete the top portion of the SWOT is rather straightforward and stems from your Skill Inventory Matrix. What are your

Strengths	Weaknesses
Opportunities	Threats

Figure 3.1   SWOT Analysis.

strengths? Those are the skills and characteristics that you identified in your Matrix. And what are your weaknesses? You can gather this data by recalling the experiences you listed in the Matrix and noting what weaknesses you had with those. They could include:

- A hard skill or skill set that you don't or didn't have.
- A soft skill on which you could improve, such as communications or conflict resolution.
- A career-advancement related skill, such as interviewing, résumé editing or negotiation.
- A trait, such as a difficulty in managing too many projects at one time or adhering to a deadline (which could also be recognized as lacking a skill in time management).

Opportunities will become clear as you engage in strategic and thoughtful networking. They will also reveal themselves as patterns of repeating Skills and Loves in the Skill Inventory Matrix and as Strengths in the SWOT Analysis. For example, if you are an electrical engineer but have always been drawn to the theatre, you will notice in your Matrix skills that relate to electrical engineering as well as theatre or set design. You will probably also see listed among your Strengths various abilities associated with these two realms. Then you could begin to research opportunities that exist at the juncture of electrical engineering and theatre.

That is what Robert Toussaint did. He was an electrical and computer science major at the UA who also loved theatre. So as an undergraduate he literally wandered over to the performing arts departments and inquired about classes he could take involving scenery and lighting design. He took more and more classes, and was doing so well that his professor made him aware of a hidden opportunity: An internship with a major scenery design firm in Las Vegas, that was to be facilitated by a UA alum. He completed the internship and was eventually hired by the company fulltime. He now works out of Vegas but regularly flies around the world assisting in the computer programming behind set movement and design for companies such as Cirque di Soleil and tours for performing artists like Taylor Swift and Pink.[i]

> **TIP:** Research potential employers and partners to discover their value and missions before you elect to work for or with them. If your value system and their value system are not in synch, this is not the opportunity for you.

Threats are anything (both negative and positive) that would prevent you from pursuing a specific opportunity. For example, let's say you wanted to work in San Francisco, but your family needs mean that you must stay in New Jersey. "Being near family in the Garden State" would be a Threat that you could list on your SWOT. This is useful information to know about your goals, just as your "Hate" list from your Matrix was useful to know about yourself. Having specific threats in mind as you pursue opportunities will also help you synthesize whether the opportunity is something you should pursue and, if so, what time frame you can and should pursue it in. Other threats could include:

- Economic, such as salary needs, the state of the economy.
- Organizational qualifiers, such as the culture of a particular company.
- Location – for example, if you want to be an astronaut, there are only a few places you can live to participate in the training. Princeton, NJ, at the moment is not one of them. So if you wanted to live in Princeton AND pursue a career as an astronaut, you couldn't do both. The location qualifier would be a threat to your professional goals.
- The values and missions of particular organizations – I touched on this above when we were completing our Skill Inventory Matrix and analyzing what we realized we loathed in our past experiences. Although your own value system is intrinsic, the value system of potential employers and partners is often a huge factor in deciding to pursue a career opportunity with that organization. If the organization for which you desire to work invests in morally reprehensible practices, this is a clear threat to the career opportunity. You can't pursue a job and mold your values to that of your employer and expect to be productive and in bliss.

> **TIP:** Self-assessment of your skills and interests will help you identify networks that you want to infiltrate and position you to land jobs and career opportunities in these new realms.

The SWOT analysis is a living document, much like the Skill Inventory Matrix. It will change and be edited as you add new experiences that you have and gain new information about yourself and opportunities from networking. But since the SWOT also incorporates external elements (the opportunities and threats), it will also adapt and change as you network more and learn about both hidden and non-hidden opportunities that you might want to pursue.

Your Career Opportunities

This book is not designed to list every single career opportunity you can consider as a result of your STEM education. Its mission is to demonstrate the power that networking has to elucidate career paths and opportunities that you may not have known existed and identify clear channels to access and pursue them. But I wanted to give you a taste of what is in store for you as you start your own career explorations via networking. Many of the following career paths I was unaware of until I networked myself. In fact, I have one of my current freelance jobs, as a columnist for APS News, the international publication of the American Physical Society (APS), as a direct result of networking. It was a hidden opportunity that I essentially fashioned myself over the course of a single phone call. This gig serves as a terrific example of accessing the Hidden Platter of Opportunities, but more to the point at hand, it afforded me the opportunity to learn about many other career opportunities for scientists and engineers. Allow me to explain.

In 2007, I was working for the UA as Director of Special Projects for the College of Science when a physicist became the university's president. I was excited about this, because having worked in physics and with physicists whom I had observed to have great leadership ability I expected that this professor would have similar leadership strengths. As I pondered his new position, I began to realize that a profile of him would make a great article. I had dabbled in freelance writing for years while I worked full-time for the UA, and I saw the physicist's presidency as an opportunity for me to pen a potentially fascinating piece (or so I thought).

But before I pitched it to any editor, I wanted to learn some more about this gentleman as well as related issues of hiring presidents in higher education. After doing a little research, I discovered that there were several physics professors across the United States who had gone on to become university presidents. Now I had a solid story pitch. I called my mentor, Alan Chodos, who at the time was the editor of APS News, with the intention of suggesting this one article. But as our conversation unfolded, another idea spontaneously popped into my head which rapidly tumbled out of my mouth: How about a column profiling physicists in non-traditional careers across the universe of industries and organizations? Alan liked it immediately and by the time I hung up from that call, I was a columnist. He named the feature "Profiles in Versatility"

TIP: If you see an opportunity, seize it. Seize it now, because it might not last!

and published anywhere from 4–6 columns each year, with each one focusing on a different physics-educated professional who had gone on to a unique career outside of academia.

I have learned a lot and gained so much from writing this column for the last eight years. I have enhanced my network, improved my interviewing and writing skills, and solidified my niche brand in the field of STEM career consulting, all of which have opened more doors to hidden opportunities and networks that I did not know existed. But if I had to encapsulate the greatest benefit that I personally received from pursuing this opportunity and writing this column, it is to make me aware of the almost dizzying array of careers that one could pursue with specifically a bachelor's degree in physics and, more generally, any degree in a STEM field. The following list gives a glimpse into the mind-blowing diversity of careers, sectors, and employment environments that potentially await you as you begin to expand your networking. And keep in mind this is only a list of careers that I have so far discovered and interviewed people in who have physics degrees. When you expand your search parameters (or decrease them to make your search much more specific) you'll be surprised by what amazing, creative opportunities lie ahead.

- Politics (elected offices), policy, and political speechwriting.
- Patent law (as a lawyer, patent agent or technology transfer professional for a university or research laboratory).
- Forensic science.
- Consumer goods: For example, a physicist who works for Proctor and Gamble as a shaving scientist and helps design blades and razors.
- Entertainment: For example, a physicist works for Pixar (which constantly hires professionals with physical science and engineering backgrounds), the creator of Futurama and co-creator of The Simpsons, and several physicists who serve as science consultants for programming such as Star Trek and The Big Bang Theory.
- Video game design.
- Global consulting, for firms like McKinsey, Booz Allen Hamilton and Boston Consulting Group.
- Museum industry, including exhibition design, curation, art conservation and restoration, outreach and public relations.
- Oil and gas industry.

- Clergy: For example, a physicist who is a rabbi and one who is a priest (and no, that is not a joke).
- Biotechnology.
- Aerospace industry.
- Commercial airline and airport industries.
- Entrepreneurship, like Elon Musk, who helped develop Paypal and launched SpaceX and Tesla.
- Ocean science.
- Sports: For example, physicists who work behind the scenes, conducting sports-related research, designing sports products, or devising the statistical methods for analyzing and putting teams together, and as athletes themselves.
- Public relations, communications, and journalism.
- Green energy industries.
- Quantitative financial analysis (the life of Quants).
- Comedy.
- Tsunami hunting.
- Extra Terrestrial Analysis: For example, a physicist who works for SETI.
- Microbrewery industry.
- Auto industry.
- Food industry (such as the guy who worked for Campbell's).
- International aid, international development, and science diplomacy.
- Archaeology.
- Art.

And by the way, how did I find the sources for all of these articles? You know the answer: I networked. I spoke with people in my own networks and used social media networking tactics (see Chapter 8) to locate professionals in these industries and careers. I also utilized networking as a way to educate myself about the very existence of these career paths, most of which I was completely unaware of.

Now, admittedly, I am privy to these assorted career tracks because I am a science careers journalist and my job (and my bliss) is to uncover and write about them. But truly, you could essentially do the same career archaeology and find and research these careers yourself using networking as one of your major tools. For example, if you were in astronomy and were interested in the entertainment industry, you could develop a networking plan that would give you access to networking nodes, such as companies, professional societies, conferences, trade publications, LinkedIn groups, and ultimately individuals who could shed light on careers in entertainment for someone with an astronomy background. And since you have superior research skills, this is not such a daunting task. It starts with speaking to people you know and inquiring if they know anyone in this field/industry/career track. Then branching out to web research and conducting a Google search for the terms "astronomy" and "entertainment" and "jobs or career." Next, take a look at the major publications that cover the business of the entertainment industry, such as Variety. Read the articles and take note of the advertisements and the industry leaders that the pieces reference.

> TIP: Networking is a very effective way of unfolding hidden career paths.

Let's look at another example. Many scientists and engineers may have heard of careers in "public policy" but may not understand what policy is, how STEM influences policy and vice versa, and the plethora of career choices that exist in the arena where science and engineering and policy intersect. I myself didn't fully comprehend the enormity and diversity of this career space until I collaborated with the American Association for the Advancement of Science (AAAS) in 2012 and 2013 to write a series of 40 articles about alumni of their prestigious Science & Technology Policy Fellowship Program. This program gives scientists and engineers the chance to be embedded in a US federal government agency or congressional office to learn firsthand how policy

> TIP: Networking and career archaeology go hand in hand.

is created and how STEM research impacts and is impacted by policy at all levels of government, as well as in companies, universities, and non-profit organizations. My job was to interview one alum from each year of the program and write a short article about them to be featured on a website celebrating the 40th anniversary milestone. With each interview, I learned more and more about policy careers and the many, many opportunities that exist in various industries for scientists and engineers who have experience in and/or an interest in policy development and implementation. In many respects, my research was a microcosmic example of networking: As I spoke with each person, they told me about opportunities they had pursued and shared with me ways to access them.

Even if I had not had this writing gig with the AAAS, I still could have used networking as a tool to conduct career archaeology; I would have started by asking my mentors, former school chums, LinkedIn contacts, and other current and former colleagues if they knew anything about or anyone in science and technology policy. I would ask to be introduced to their friends so I could discuss their experiences with them. Additionally, since networking is not just about uncovering information that is useful to me but also revealing information to the other party to help them in some way, I would also endeavor to find out how I could potentially assist them in their work presently. I would additionally conduct web searches using key terms like "science and policy" and see what people, organizations, websites, publications, articles, videos, and so on popped up. And I would leverage all of those search results as a channel to connect with the people mentioned in those articles, and so forth, to conduct informational interviews and develop a true networking relationship with them.

The key point is this: Whether you already know what profession will bring your bliss, or you are still analyzing your own interests and goals to see how they synergize in a career path that is right for you, networking can help you find resources to make intelligent decisions about advancing your career in the direction you choose. Take advantage of your networks, networking nodes, and your networking abilities to educate yourself about the many, many jobs and careers that are available to you with a STEM education.

Chapter Takeaways

- As a STEM-educated professional, you have seemingly endless career options.
- Your major professional goal should be to find a career that provides you with bliss.
- You will know you are in bliss when the activity you are pursuing is intrinsically rewarding and strikes a balance between undertaking a high degree of challenge while using a high level of skill.
- You can utilize various self-assessment tools such as a SWOT Analysis to help you identify career paths you want to pursue that align with your interests and goals.
- Networking will be one of your most powerful tools for conducting Career Archaeology – discovering career paths you didn't even know existed.
- Networking can help you find resources to make intelligent decisions about advancing your career in the direction you choose.

Note

i.   Levine, Dream Jobs 2012: Designing Automation for Acrobats, *IEEE Spectrum*, January 31, 2012, http://spectrum.ieee.org/geek-life/profiles/dream-job-2012-designing-automation-for-acrobats.

4 Establishing Your Brand and Reputation to Gain Access to the Hidden Platter of Opportunities™

Once you know what your brand is, you need to engage in activities that allow for the communication and promotion of your brand to strategic publics who know about, can conceive, and can give you access to hidden opportunities.

In this chapter, we delve into specific strategies that allow you to communicate your brand to decision-makers. The strategies and tactics contained below are all significant elements of networking. But it all starts with asking questions.

Always Ask Questions

Why do we ask questions? What does it signal to the party to whom we address our questions? It demonstrates that we are hungry for knowledge and thirsty for information. It shows we want to improve ourselves and our skills and to seek out new means for solving problems. In sum, when you ask a question, you establish that you are a success: Successful professionals continuously look for opportunities to advance their trade and knowledge base so they can improve their skills, expand their problem-solving abilities, increase their productivity, improve and advance their organization, and ultimately experience bliss in the process. Successful people remain successful by remaining inquisitive.

> TIP: People who are educated and are successes remain so by continuously asking questions and seeking to improve; those who don't, keep their mouths shut and stay exactly where they are now with no change.

Others respond positively to those who ask questions. They see you as someone who is passionate and enthusiastic about something, whether it

Networking for Nerds: Find, Access and Land Hidden Game-Changing Career Opportunities Everywhere, First Edition. Alaina G. Levine.
© 2015 John Wiley & Sons, Inc. Published 2015 by John Wiley & Sons, Inc.

is your field or someone else's, and they look for ways to not only answer your inquiries, but to also delve more into the subject themselves. In fact,

> **TIP:** Champion networking relies on an exchange of questions and answers between two parties. It delivers ROI to everyone involved.

being inquisitive is considered such an important quality, especially in the sciences and engineering, that when you demonstrate this, you solidify and amplify this aspect of your brand and attitude in the minds of those you are engaging. And the more the other parties know about your brand, the more they will reveal hidden opportunities to you. In sum, champion networking (that delivers ROI to everyone) relies on an exchange of questions and answers between two parties. So as you plan your career strategy and networking plan, get into the habit of asking questions and demonstrating your keen desire to know more about the world.

Now of course part of the reason you want to ask questions is so that you can learn as much as you can about your field. You want to be the best you possibly can in your area of expertise and one of the most straightforward routes to achieving this is to engage other experts and ask them about their work and problem-solving tactics. So when you have the opportunity to speak with other disciplinary leaders, do so, even if it's just for a few minutes and even if it's just to ask a few questions. I have known people who ran into someone they really wanted to meet, at a conference for example, and had literally five minutes to spend with the other party. They used the time to ask and answer questions and it led to fruitful collaborations, grants, and co-authored papers.

But asking questions of experts goes beyond just the technical discipline and can include everything that goes along with it, such as the business side of being a scientist or engineer, the culture, and the manners in which scientific problem-solving is conducted.

One of your goals is to become a leader in your field, and a leader does not necessarily imply seniority in a discipline or having a managerial position in a workplace scenario – it suggests excellence, dedication, significant contributions to the arena, and entrepreneurial problem-solving. Leadership in a field signals that you set the tone or a standard of quality as well as quantity

> **TIP:** Being a "leader" does not necessarily imply that you are a senior member of your organization or profession, or that you are managing people or projects. A leader is someone who has the vision and ability to achieve organizational and/or professional goals and enables the success of those around them.

and that others see you as a trend-maker. It means that you have vision and goals and can design and implement strategy to achieve those goals. And finally, leadership denotes the ability to inspire and enable the success of those around you.

If you have completed a doctorate or even a Master's thesis, chances are you are the only person on Earth who knows as much as you do about this microsubject. That automatically makes you a leader in your particular subfield. I don't want you to think of this as a daunting concern. It doesn't make you better than anyone else and you certainly shouldn't articulate your expertise with an underlying thought that indicates you are snobbish, entitled, or a know-it-all. Rather, I want to point out a fact: The dissertation or thesis you completed was original research. The knowledge about the subject did not exist before you researched it and brought it to light. Therefore, you and only you understand the depth and complexity of this sub-sub-area more than anyone else. Of course there are experts who know more about the overarching field and even sub-field. However, if they have questions about your sub-sub-discipline, you are the professional they will go to. Your first-hand, intimate, and uniquely-held knowledge serves as the basis of your leadership in your sub-sub-field.

> **TIP:** Asking questions about your field can improve your leadership abilities and brand you as a leader.

But your objective is to expand that leadership to encompass other areas of your discipline, to master the skills associated with being a leader, and then to brand yourself a leader. The reason? It's simple: When you are perceived as a leader, others come to you and offer you access to the Hidden Platter of Opportunities. So the more you ask questions of other leaders and observe their leadership styles and contributions, the sharper your own leadership skills will become.

> **TIP:** When you are perceived as a leader, others will come to you to engage you and offer you access to the Hidden Platter of Opportunities.

Asking Questions of non-STEM Professionals

Not only do you want to become more of an expert and leader in your own field, but you also want to gain information about other fields, arenas, and even sectors so you can do your job better and plan your career path with more enlightenment.

So seek out people who are not necessarily in science or engineering and ask them about their work.

The reasoning behind this is easy to understand – you never know what information you are going to get from someone until you engage them. And as I wrote above, your aim is for a diversity of ideas and sources of information and inspiration. Companies, organizations, and indeed individuals must continuously be innovative if they are to advance. They must constantly look for new opportunities, new problems to solve, and novel ways to solve them. If they don't, they become stagnant and risk losing out – either losing a connection, an opportunity, or even a job.

> TIP: You really never know what information you are going to get until you engage someone.

So how do you stay on top of innovation? How do you ensure that exercising your abilities to think and solve problems innovatively is part of your natural day-to-day activity? You seek out and engage a diversity of sources who can provide you with a diversity of ideas which leads to innovation.

This diversity of ideas concept is absolutely necessary as you network. So if given the chance to speak with someone (like on an airplane) who works in a job that seems foreign to you, don't immediately dismiss them and refuse to speak with them. They could provide you with invaluable information that could change the course of your career.

My favorite story of the benefits of asking questions comes from my uncle. Here was a man with excellent pedigree who also happened to be a genius. He was an applied mathematician by education whose undergraduate degree was from MIT and PhD was from Harvard, and he had been a successful software entrepreneur. But despite his credentials and intelligence (or perhaps because of it), this nerd would often ask the garbage man who stopped in front of his house about his work. With great respect, he would inquire about everything from how the hydraulics on the truck operated, to what materials were used to construct the cans, to how they determined and mapped out the schedule of pick-ups. Was this

> TIP: By asking questions of people who are masters at their own craft, you can learn something that can aid you in becoming a master at your craft.

prosperous scientist and entrepreneur interested in transitioning into the trash trade? Absolutely not. But what he recognized stays with me even to this day: When you engage someone who is a master at their craft you can learn much to improve your own craft.

And the way I like to think about this is to assume that one day the garbage man gave my uncle a visual image of how the hydraulics on the truck function. My uncle then stored that information in his mind and didn't access it for weeks, months or even years. But perhaps one day he was working on a piece of code that was caught in an infinite feedback loop and suddenly he remembered the image of the hydraulics.

He applied that knowledge to his own software problem and was able to come to a favorable solution. This is a prime example of the power of asking people questions and engaging them in conversation. You really never know what information you are going to get, and you can almost always glean at least a nugget of pertinent value that can help you sharpen your own talents and solve problems in your own field, even if that person is not in your discipline or industry.

There have been many, many times in my life (too many to count actually) where I have engaged a stranger in conversation, asking questions about their own life and profession, only to find out strategic information which has changed the course of my career. An important tactic to utilize when having a conversation is to ask questions about the other person. Ask them about the best part of their job or what brings them bliss, or how their passion is manifested in the work they do. As you probably know, people love talking about themselves, so the more you engage them in conversation about them and what brings them joy, the more they will open up to you and look at you as an ally.

> TIP: Get people talking about themselves, a subject most find great enjoyment in!

Furthermore, when you have a discussion with someone about ideas and actions and activities that bring them happiness, they are more likely to remember you when you follow up with them after the initial point of contact. If you and I meet at a cocktail party and you ask me about science writing and why I find it so exciting and fun, my eyes are going to light up because I am going to be concentrating on something that brings me pleasure. And when you follow up with me via email and remind me that we had a conversation about the pleasures of being a science journalist, I am going to equate you with the mental note of why I love my job. I therefore am going to be more apt to speak with you about it.

> TIP: Get people talking about what brings them pleasure.

Consider the opposite scenario, in which you engage someone in discussion about something that they find unpleasurable, and you say something to them that evokes a hostile thought or memory. If you meet me and after learning I am a science writer, you exclaim "Wow! That's horrible! That must be so difficult and challenging, especially in this economy! Why would you pursue *that* job?", imagine how I might respond. I immediately will feel uncomfortable. My brain will immediately wander over to negative ideas, perhaps about my vocational choices. But more importantly for you, I will immediately shut down. I won't want to

> TIP: The most effective networking authentically seeks to elucidate information for both parties that allows you both to make decisions to advance your careers, projects, and organizations.

speak with you anymore and any notion of the Hidden Platter of Opportunities will be withdrawn. Who wants to be reminded of something horrible, or feel that their career choice is being insulted?

So stay on the sunny side of networking and asking questions! And it is important to note here that I am not advocating anything sleazy; in fact, I am suggesting that you be completely authentic in your inquisitiveness. In the example above, you want to know why I like science writing, so you ask me about it in a way that reminds me of my pleasure from it AND gives you strategic information about the field AND positions us both to build a mutually-beneficial partnership. This is effective networking and it is also extremely authentic – it provides us both with information to be able to make decisions to move our careers, programs, and projects forward.

> **TIP:** When networking, especially with someone who you just met, you should be listening most of the time and talking very little. But when you do talk, ask questions and insert information about your own experience that's relevant (see Chapter 5 on Informational Interviews).

And although knowing the negative attributes of a career or employment option is strategic information for you to have, you don't need to bring this up in the first conversation. The initial point of contact, where we begin our networking partnership, should focus on positive elements of our trade. As we build our relationship and as we get to know each other more, there will be future moments in time when it is perfectly appropriate for you to ask about some of the undesirable aspects of a job, career, organization or even a region. But just like on a first date you wouldn't talk trash about your ex-girlfriend, or ask me about my horrible divorce, on the first networking exchange, keep it positive!

When I began what I refer to as my hard-core networking, which was in the early 2000s, I asked a lot of questions of people with whom I met. I remember one particular experience that nicely illustrates my point about the value of asking questions. When I started my job in the as Director of Special Projects reporting directly to the Dean of the UA College of Science, my responsibilities included overseeing the brand new Professional Science Master's (PSM) Degree Program in Science and Business. This novel program was designed specifically for science students who wanted to work in industry, so naturally part of my job was helping my pupils connect with industrial decision-makers to find employment. As a neophyte to the industrial landscape, I knew I needed to immediately expand my network to gain external, business-based partners and advocates, and to promote the program's benefits to potential employers and economic development leaders. And since it was important to find allies close to the UA, so students could also pursue internships while taking coursework, I started my networking by sending out a number of cold emails to companies and organizations in Phoenix, the state capital, which was about 100 miles from the UA.

On one Friday in early December I drove up to the "Valley of the Sun" to have a series of meetings with potential industry partners. My last get-together of the day took place in the penthouse suite of one of the tallest buildings downtown. The office overlooked the entire valley. Needless to say I was impressed. As we were concluding our discussion, I couldn't help but notice that there was a bit of a commotion taking place in the hallways; it turns out, the organization was preparing for its annual holiday party.

I had never met the gentleman with whom I was conversing before that day – I had sent him an email out of the blue and asked for the chance to chat with him about the UA programs. So he wasn't my friend and he wasn't my associate – yet. But as I saw them setting up for the gala, I knew I had to take advantage of this opportunity. Knowing his organization, I immediately realized that the networking potential at this affair was going to be fabulous – the organization was a star in the community and industrial arena of the city and state, and thus would naturally attract leaders from other high-quality firms who also would recognize the networking importance of attending. The party itself was poised to be a rich networking node.

> **TIP:** Take note of how people interact with you at networking functions. If they are rude, unprofessional or downright offensive, this gives you valuable information that you do not want to collaborate with them!

So as we prepared to say our good-byes and shook hands, I opened my big mouth and I said, "I notice that you are setting up for your holiday party. Do you mind if I join you?" Now, what was the worst that could have happened? This was a very calculated risk I was taking, but it was a very low risk. The worst possible thing that could have occurred would have been him saying "no, unfortunately we don't have the room" or the food or whatever or he might not have given an excuse and just said no. And that would have been fine. It probably would not have affected any future collaboration that could have transpired. (Of course if this had taken place in another country, the cultural norms might have dictated a dif-

> **TIP:** If you spot an opportunity, it (usually) never hurts to ask!

ferent scenario for both me and the other guy. In some cultures, it would have been rude for me to ask.) But in this case, he most certainly wouldn't have said "Absolutely not! How DARE you invite yourself to my exclusive party! Be gone with you, you contemptible, rude cow!" (Of course if he had remarked in this manner, it would have spoken volumes about him and his organization and given me reason to terminate any possibility of partnering with him.)

> **TIP:** Professionals, especially those in managerial roles, look for talent who distinguish themselves by asking for what they need.

But in this case, the businessman answered in the affirmative. "Sure!" he said. "Why not? Come on in and enjoy the party," and before I knew it I felt like the belle of the ball.

So I grabbed a drink and some finger food and wandered out to the sprawling veranda to admire the sunset. And as I was there, feeling mighty proud of myself that I had gotten access to what appeared to be an

> **TIP:** You don't need an opening line at networking receptions! Just introduce yourself.

exclusive party of the Who's Who of the Phoenix business community, a man drifted over to me. Not being the expert, suave networker I am today, I thought I needed an opening line to start a conversation. So I pointed to the mountains in the distance and stylishly let slip, "nice view, eh?"

He responded positively, smiled, and that's when I felt comfortable enough to extend my hand and introduce myself. "I'm Alaina Levine with the University of Arizona," I said. He answered: "Hi, I'm John Smith." He didn't state his affiliation, but I could see his nametag had printed "The Honorable John Smith." And since I was new to networking, I couldn't think of anything else to say except "Are you the mayor of Phoenix?" because I wrongly assumed that if he has a title like that he's got to be the mayor. Of course that was incorrect – there are lots of job titles that come with "The Honorable," especially in state and regional government. As I soon found out, this guy was a state senator. But I recovered well from this potential faux pas, because I immediately jumped to a subject I knew he would enjoy chatting about: Himself. I delightfully exclaimed "Wow! That's terrific! You must have a lot of satisfaction helping people. What's the best part of your job?" He immediately grinned ear to ear and spent the next 20 minutes expounding on all of the wonderful and meaningful things he did and had done for the state of Arizona.

That was the first time I realized I didn't need an opening line. It was also the first time I realized that when you network, you should think about the other person first, and ask them positive questions that get them talking about themselves and their bliss.

Stupid Questions

I have often been asked (ironically) if there is such a thing as a stupid question. I once gave a talk about succeeding in college to a bunch of middle school kids and mentioned that stupid questions do not exist, like the Yeti or dinosaurs hunting with humans. A young boy raised his hand and said "Well my teacher says there is such a thing as a stupid question!", to which I replied: "Well clearly, your teacher is stupid."

TIP: There's no such thing as a stupid question.

Why on Earth would I say such a thing? Why would I jump to the conclusion that the teacher was stupid for thinking that some questions are stupid? You know the answer: Because no question is *ever* stupid. When you ask a question you are signaling to another that you aim to become more educated, more learned, and more knowledgeable about your field and the universe. Educated people continuously ask questions because they want to improve their skills. So don't ever think your question is stupid.

I would argue, however, that there is such a thing as an appropriate time and place to ask a question, which is entirely relative to your comfort level as well as the cultural norms of the ecosystem in which you are making the inquiry. For example, let's say your background is organic chemistry and you find yourself in a lecture being given by astrophysicist Stephen Hawking. The audience is filled with astrophysicists and for some reason, you either don't know or forgot what a black hole is. In the middle of this sea of experts, it might not be appropriate for you to raise your hand, interrupt the good doctor, and ask a question about a fundamental aspect of astrophysics. However, if after the address is completed, you go up to Professor Hawking and introduce yourself, and tell him how much you enjoyed the talk and ask where you can get more information about black holes, that would be entirely appropriate (and most likely welcomed by the speaker as well).

As you become more adept at and comfortable with networking, you will also become more savvy in your ability to judge whether you should ask a question in a particular time and place. This gives you one more reason to pursue opportunities where you can conduct low-stakes networking, like on an airplane or at the grocery store, so you can begin to practice your question-asking and become more astute at recognizing your own comfort level based on the individual scenario.

Seeking Mentors: The Importance of Mentor–Protégé Partnerships in Networking

As you network, you will find that there are people who almost naturally want to be closer to you, more involved in your work, and more helpful to you. At the same time, you will also be organically attracted to them and will innately find ways to collaborate and help each other. These are the people who will be your mentors. And building and leveraging the mentor–protégé relationship is an extremely critical component of successful networking.

I used to think that a mentor was someone like an advisor in college, who could provide me with "advice" relating to classes, internships, graduate school, and jobs. Their function was to dispense this advice like a doctor dispenses prescriptions and tells me what to do if my knee hurts. If I want to study topology, my mentor would tell me what graduate school had the best topology program. Additionally, their role was also to write letters of recommendation on my behalf to pursue these opportunities.

> **TIP:** The mentor–protégé partnership is only as strong as the value that both parties provide to the other.

But this description of a mentor barely scratches the surface of what a mentor's true value is, and the invaluable nature of the win-win partnership you craft with your mentor. Many early-career scientists and engineers erroneously believe that the affiliation with their mentor is a one-way street – that the mentor provides all the value in the equation. But in actuality, just as is the case with every partnership you build from networking, the mentor–protégé partnership is only as strong as the value that both parties provide to the other. It is a symbiotic relationship and it can set you and your mentor up for success now and down the road.

Here are some key thoughts about mentors:

- A mentor can be someone as formal as a boss or Principal Investigator, or he/she can be someone who is in an informal capacity. Perhaps you met them at a conference and stayed in touch. Perhaps you worked with them on a short-term project or they got to know you from serving on the same committee. No matter how they found out about you, the key is that they did so because you effectively and appropriately promoted yourself and your value to this person. I have mentors from various phases of my many careers, some have been my supervisors and other relationships were formed organically – I was attracted to them and their work and vice versa.
- A mentor helps you because they get something out of it. Just like all of your networking relationships, the mentor expects something of value from the partnership. This value may not materialize for a while, but it's there – they are not mentoring you out of a philanthropic urge. They do want to help the next generation of science and engineering leaders, but they also want to help themselves, as they should because this is meant to be a mutually beneficial partnership.
- A mentor wants to help a star. Mentors know what they are looking for in protégés – they are seeking collaborations with professionals who are intelligent, ambitious, passionate, hard-working, productive, deliver on their brand, and have a positive attitude. In short, mentors want to mentor stars. Your attitude goes a long way in showing others that you are a star. And by star I don't mean a cocky jerk. I mean

someone who wants to improve themselves, those around them, and scholarship itself. Mentors positively respond to people like that and want to help them.

- A mentor has access to networks you don't have access to. This is a concept that was first introduced to me by Howard G. Adams, a professional speaker and author whose expertise lies in fostering mentor–protégé relationships. Up until I heard him speak at a conference, I thought of a mentor as simply a conduit of information. But Adams expanded on that definition and clarified what a mentor can really do – their networks are priceless and they have access to people, places, ideas, and data that you don't have. As such, they can be almost indispensable when it comes to expanding your own network and gaining access to the Hidden Platter of Opportunities.

> TIP: "A mentor has access to networks you don't have access to" – Howard G. Adams.

Think about this – let's say you are my protégé. I have access to people you might want to meet, people who could change the course of your career. And let's say as a result of you being my protégé I have gotten to know you and your work, your strengths, and your ambitions. Your expertise is in engineering computer-based systems and one of your goals is to work for Microsoft. And then one day I attend a party with Bill Gates. I have access to this party and to Gates, but you don't. As I chat with my buddy Bill, he shares that he is currently seeking someone with a background in computer engineering with a proficiency in engineering computer-based systems and laments that it is so difficult to find quality hires these days. As your mentor, I can swoop in right there and give him your resumé or introduce you via email. He knows me, so he knows that I

> TIP: Your mentor can provide invaluable assistance in finding people with whom to network or identifying and accessing specific networking nodes that will help your career.

can vouch for you. I have solved his problem by finding him a qualified candidate for the job, and I have solved your problem of helping you land your dream job by giving you access to this hidden opportunity.

- Mentors often act like matchmakers. If they know of two people who have outstanding brands, attitudes, and reputations and who share the same values or passions, and who they think would make great partners themselves, the mentor will often introduce the two of you without you even having to ask. This happened to me with one of my mentors, a business leader in Southern Arizona. After talking informally with him for just a few months, discussing mutual interests and our background, he suggested that I meet another one of his protégés, who happened to be a rising star in politics in the region. As a result of his matchmaking, I was able to connect with this person and assist her in her work and vice versa.

- You can learn from your mentor how to be a mentor. Of course as you advance in your career, you'll find yourself in scenarios where you are the mentor. The earlier you learn how to be an effective and helpful mentor, the better it will be for you and your protégés. And of course as you learn to be a successful mentor from your own mentors, you pay it forward to your protégés and then you even pay it back to your mentor. It's a truly wonderful cycle that delivers excellent networking and career development and planning ROI.

- You can help them too. Don't think that because you are at a certain phase in your career that appears to be junior to your mentor that you can't help them. You can always assist them – in very similar ways that they assist you – by being resourceful and giving them access to networks, ideas, and inspiration that assist them in their endeavors. But most early career STEM professionals tend to think of themselves as the "lowly" protégé in the equation, when it really is supposed to be a win-win relationship. In addition, as I wrote above in Chapter 1, the mentor values the relationship with the junior protégé because in many respects you are their legacy – you are the professional who will carry their work into the next generation, expanding on it, bringing it to new heights and serving humanity with it even better than they are. So this is a good thing to remember as a protégé.

> **TIP:** A successful mentor–protégé relationship delivers value to both parties. Even as a protégé, and no matter where you are in your career, you are providing something important to the mentor.

I received confirmation about this issue of being a legacy to a mentor when in 2012 I had the great fortune of winning a travel grant to attend the 62nd Lindau Nobel Laureate Meetings in Lindau, Germany. This is an annual event centered on a different subject in which the Nobels are given each year. The summer that I traveled to Germany, the conference focused on physics. Approximately 30 Laureates and 500 students and postdocs from all over the world attended the event. The students who have gone call it "life-changing" and I agree – the Laureates gave lectures in the morning and then met in small groups with the youngsters in the afternoon. There was plenty of opportunity to network with other scientists from around the world as well as the Prize winners themselves, who often had lunch and dinner with the students. I was there as press; and as a member of the media, I interviewed a number of physics

Laureates about their experiences at Lindau and what kept them coming back year after year. The answer was pretty much the same for all of the gentlemen with whom I spoke – the early-career scientists and engineers who were there were the *legacy* of the Laureates – they would transport their work to the next generation into the next decades, perhaps even the next century, they would be the ones who would solve the problems that stymied the Laureates their entire careers. So the Laureates honored the students as a needed element of furthering scholarship. That is the same mentality that drives profitable mentor–protégé partnerships and is the foundation for strategic networking.

- Think long-term. The great thing about a true mentor, a professional who really wants to help you with your career and see you achieve your dreams, is that you don't have to constantly speak with them. You can apprise them of accomplishments, send them inquiries about topics as they come up, and contact them about their contributions, but you don't have communicate with them every day. Just as with networking, where the partnership only ends when one or both of you drop dead, so too does the mentor–protégé relationship. They can be mentors for life. I speak with several of my mentors only a few times a year and each time it is a fruitful experience. They can help you throughout your life, and you don't have to rush the relationship.
- Strive for diversity with your mentors. I have mentors who are scientists, journalists, and engineers. I also have mentors who are in business, policy and politics, and law, and mentors who are tradesmen, technicians, and clerks. The value of a mentor is not necessarily defined by their vocation – it is determined by their own success and knowledge and their interest in helping you succeed in your field, even if it is different from theirs. And as I keep emphasizing, just as diversity is the name of the game in networking, so too is diversity crucial in forging mentor–protégé partnerships. A diverse set of mentors from varied sectors can give you advice and access to networks that are unavailable to those who only have mentors in one realm.

TIP: If someone asks you to stay connected with them or asks you to follow up with them, they usually mean it. Do so!

One of my closest mentors is not even remotely in my discipline, but has helped me in more ways than I could have envisioned. I met him at a speech. In 2001, I gave a talk about the UA College of Science to a local group of business and economic development advisors in Tucson. It was designed to share information about our programs and the hidden value we provided the community, and why the leaders might want to partner with us. After my presentation, which was probably only 20 minutes long, one gentleman ran to the front of the room to introduce himself to me. He gave me his card and told me that he enjoyed my speech and would like to learn more. After the meeting, I sent him an email and arranged to have coffee with him. It turned out that this man was a very powerful and well-connected industry leader in Tucson. He expressed that he was impressed with me and my attitude and wanted to stay connected.

This man soon became more than a contact for me. He became a mentor and a source of inspiration with whom I could bandy ideas about. He encouraged me to analyze and solve my day-to-day problems in novel ways, and helped me advance my career in new directions I probably would have never considered. And he introduced me to new people and aided me in expanding my networks into new directions I would have never been able to access on my own.

Now to be clear, this man was not in science and engineering. In fact his business was real estate mergers and acquisitions (M&A). I happen to hate the real estate business (although I didn't tell him that at our first meeting!) but that didn't stop me from pursuing a rich relationship with this professional. As I mentioned above, he was a master craftsman, and even if it was a craft I would rather stick a fork in my eyeball than pursue, I knew he was able to provide me with ideas and information to assist me in my craft. In addition, I recognized that this person was a seasoned professional and had knowledge of other aspects of the business ecosystem which could be very helpful to me in my career. And in fact, he helped me with a project in just that way.

One afternoon, I was invited to be interviewed on a radio show about the PSM Programs in Applied Science and Business that I was overseeing at the UA. Now I was a radio novice; the extent of my experience on the radio consisted of me being the 5th caller and screaming into the phone "did I win the Duran Duran tickets?" This was a big opportunity for me, the program, and the school to get some great PR and I didn't want to blow it. I didn't have a contact in the radio biz from which to seek advice, so I called my mentor, the real estate M&A guy. I knew that even though he wasn't in communications, he probably had experience in giving radio interviews and had some insight into how to approach and perform this task well, to optimize any PR ROI. As my mentor, someone who was watching out for my back, I knew he would be able to guide me to ensure I did an excellent job.

But of course there was another explicit reason that I wanted to ask his advice. I wanted to share my achievement with him and promote my success. As I discuss below, networking is inherently linked to self-promotion. But the key is to engage in appropriate self-promotion activities. By calling up my mentor and asking his advice on this radio interview matter, I was promoting my PR skills to him as well as the fact that a third party (in this case the radio producer) thought my work was worthy and interesting enough to put me on the radio.

So when I reached out to my mentor, we spoke for only a few minutes as he gave me counsel. I jotted down some notes, and thanked him for his time and for sharing his expertise. I then sent him an email containing the information about when the interview would appear on the radio, so he could listen.

The advice was fantastic – I entered the situation with much more confidence than if I hadn't discussed it with my mentor. He aided me in my professional output – the radio interview – despite the fact that he wasn't in radio or in science or even in communications. And of course, following the interview, I sent him a hand-written thank you note to express my appreciation for helping me achieve what I considered a successful milestone in my job and career.

> TIP: The mentor–protégé relationship doesn't mean that you both have to be communicating with each other all the time. In fact, a strong partnership can stand the tests of time and changes in geographic locations and even jobs and industries.

My relationship with this individual has lasted more than a decade, and in that time, he has provided me great value and I have helped him as well. We help each other, which is why it has remained fruitful. We may not see each other or even speak very often, but I know that if I had a question or concern or an idea to help him in his business, he would be forthcoming and would welcome the continued interaction.

Now we would hope that all mentor–protégé alliances are always rosy and friendly, but people are human after all, and there may be situations where you and your mentor (or protégé) get into a tiff. In addition, just like with dating, not every potential mentor is going to want to work with you. Your personalities and values might clash or "he just might not be into you." So how to you handle the delicate and sometime confusing landscape of mentor–protégé interactions that aren't so sunny? Here are some thoughts:

- Resolve potential conflicts or miscommunications quickly. You don't want to let anything fester with a mentor, so if there is a problem, address it as soon as possible.
- Not everyone will want to be your mentor: It is important to be honest about the way the world works. Not every person who you would like as a "mentor" will want to work with you or for that matter be in the same room with you. This is just a fact of life and you don't have to worry about it too much because there are plenty of other people who will value you for you. I know this sounds a little like dating and in a certain respect it is very much like that. If you are pursuing someone who you would like to be your mentor, and after ten emails and two years they haven't responded to your notices, it is now no longer an

issue of them being "super busy." They are ignoring you because "they are just not that into you." And that's ok. Let them go and you get on to the next partnership which could produce fruitful returns for all parties.

- Finally, know who your true mentors are and who might be an enemy or, what I refer to as, a professional zombie. The Godfather said it best: "Keep your friends close and your enemies closer." It is critically important that you be ensure that your "mentor," the person with whom you are collaborating, and on whom you are relying for potentially life-altering information, is actually your mentor and not an enemy. The last thing you want is to have someone in your life who appears to be helping you but is actually a zombie – working to take you down, ruin your reputation and suck the creative life out of you. This is a big concern for early-career scientists and engineers who are doing exceptionally well. Not everyone who sees a star wants to help make that star brighter – sometimes they just want to suck the brightness and creativity from that star and use it themselves. So you have to watch out for zombies, especially early in your career, when your reputation and brand are at their most vulnerable.[i]

In your role as a protégé:

- Learn the tools, tips, and tactics of your trade, so you can be even more successful at your job, in contributing to scholarship, and to assisting the mentor when needed.
- Heed their advice: Mentors will often give you advice whether you ask for it or not, directly and indirectly. They might suggest you apply for an award, which might be code for "apply for this award because there is a big chance you will get it since I know everyone on the committee." They may give you non-verbal cues as to certain directions you are taking or decisions you are making. For example, if you are on the phone or in person with the mentor and you are discussing a certain scholar with whom you are hoping to do a postdoc, you can gain insight as to whether this person would be a good PI by observing your mentor's body language and dramatic pauses. So learn the communication styles of your mentors so you can gain clandestine information (direct and indirect, verbal and non-verbal) that they don't reveal up front.
- Look for opportunities where you can assist your mentor: If you see a chance to co-write a grant proposal, or if you hear about a job that you think your mentor would be perfect for, tell him. If you learn of an award that your mentor would be right for, ask him if he would be comfortable with you nominating him. Look for opportunities to assist them with their day-to-day work, their networking goals, and self-promotional aims. They will appreciate you did this and your reward will be an even more enriching relationship.

- Look for opportunities to practice mentoring others: No matter what job or career you pursue or sector in which you desire to work, there are always opportunities to mentor others. This is a skill that employers prize, so take advantage of it – it demonstrates you know how to oversee subordinates, lead teams, and resolve conflicts. And of course you want to pay it forward …

The Bottom Line: You will have many, many mentors and protégés in your career. Each one of these relationships is an invaluable element of career advancement, and each one provides precious opportunities for networking, identifying relevant networking nodes, and accessing the Hidden Platter of Opportunities.

Being Professional

> **TIP:** A professional is simply someone who is serious about their craft, and every action they take solidifies and amplifies that dedication.

Your role as a professional is intimately tied to your brand, attitude, and reputation and thus plays a huge part in expanding your networks. But what exactly is a "professional"?

The definition of a professional is simple: It is someone who is serious about their craft and exhibits that seriousness in every possible way, including:

- How they interact with others.
- What they say.
- How they say it – their use of vocabulary, tone, and poise.
- Their attitude.
- How they approach their work.
- Their attention to appropriate and culturally-mandated etiquette (in interactions, meals, and correspondence).
- What they wear and when they wear it.

The importance of being professional cannot be overstated. If you are perceived as a professional you will be treated as a professional. This serves to elevate your brand and amplify your reputation. It is a critical lynchpin in networking, because if people see you as a

> **TIP:** Being known and seen as a professional is especially vital in networking, because if people view you as a professional they are more likely to engage you for a career opportunity, whether it is hidden or advertised.

professional they are more likely to engage you for a career opportunity. Think about it this way – the opposite of a professional is someone who is an amateur, hobbyist or enthusiast. They consider their work something enjoyable but not something that demands their full attention or that takes priority in their lives. But you, my friend, are a *professional*, which means that I know you are so serious about your craft that you will stop at nothing to solve the problems associated with your craft. Professionals want to engage other professionals for employment and other significant collaborations.

One of my favorite stories about how critical it is to establish yourself as a professional from the first moment of contact stems from a time when I was on a search committee for a new employee in my department. The person was interviewing for the position of development director – in other words, she would be the public face of the unit as she endeavored to raise money on its behalf. If ever there was a job that demanded a level of professional decorum, it is the chief fundraiser.

Considering that it was an influential position, the candidate partic-ipated in multiple interviews with key campus leaders throughout the day. My committee was scheduled to interview her around 4pm. Now of course, even though it was at the end of a long, tiring day, it would have been strategic for her to maintain her energy level throughout the expe-rience and demonstrate a continuously positive and professional attitude no matter how many people she had to meet during the hiring process. She needed to make an all-round good impression with every interviewer if she were to land the job.

But when I entered the interview room a few minutes before her appointed time, I looked around and at first did not notice her. I saw a woman in a suit, but she was chewing gum, twirling her eyeglasses in her hands and shifting back and forth from the balls of her feet to her ankles, like a little kid does in the playground. Little did I know that this was the person who was auditioning for the role of fundraiser. She had been on a break between two interview committees, and in full view of everyone was hanging out like she was at a party. She didn't acknowledge my teammates and I as we walked in but instead continued snapping her gum and chatting with some other gal.

When the interview was formally called to order by the chair, I finally realized that this woman who had been acting extremely unprofession-ally was the candidate. But her unprofessionalism did not end when we commenced the interview. As we asked her questions, she still continued to chew her gum and twirl those darn glasses. Needless to say it was very annoying and almost shocking – you would have expected to see some-one display a little more professionalism in an important job interview for a position that demanded 100% professionalism100% of the time.

But then it got even worse – I asked her if I could see her portfolio which she had brought with her, and instead of passing it down to me, or getting up and dropping it off for me and then returning to her seat, she walked over and proceeded to stand over me and turn the pages for me and stick her finger in my personal space as she endeavored to point out some of her proudest accomplishments in the book.

Needless to say this person didn't get the job. In fact, her attitude, demeanor, and actions solicited snickers from me and the team immediately following the incident, and even for years later. And what we laughed at was this: She should have known what it means to be a professional development director. She should have known the culture of our institution and the cultural requirements to successfully serve as a fundraiser for a major university. She should have known that simple actions like popping your gum or physically invading someone's personal space during an interview established her as the complete opposite of a professional. She should have known that we would make snap decisions about her ability to be a champion fundraiser based on her unprofessional behavior.

> **TIP:** Your actions and manners speak volumes about your brand.

And that's the way the world is – people do judge you on your attitude, as I wrote above, and they judge you on your professionalism. Being professional and being perceived as a professional go a long way in solidifying networking alliances and gaining access to hidden opportunities. So here are some key thoughts about the subject:

- Act like a professional and you will be perceived and treated like one. Your actions and manners speak volumes about your brand. You teach people how to treat you by your actions, so make sure they only see you as a professional and thus treat you as one.
- Being professional involves adhering to certain norms and standards of etiquette for the environment and culture in which you desire to work. Practicing proper etiquette, in networking and other interactions, at meals, and in correspondence (including via telephone, email or on social media), demonstrates your professionalism and your commitment and dedication to your trade and thus your attitude and brand (see below for details about proper etiquette).
- Never lose your cool: As a science or engineering professional, you have a job to do that involves very specific problem-solving. Your only goal, if you want to keep your job and advance in your career, is to keep your eye on the ball and keep endeavoring to solve the problem. Leave emotion at the door. Just as Tom Hanks said in *A League of Their Own*, "There's no crying in baseball," there's no crying or screaming or pouting in STEM.
- Maintain your professionalism, no matter what anyone else does. People around you may behave differently, but you stand strong. There are always going to be others who don't know what professionalism in their field entails, or eschew it altogether for their own

personal reasons. But you have to always remain professional because your reputation and access to the Hidden Platter of Opportunities absolutely depends upon it.

I often notice unprofessional behavior in places where it shouldn't be happening, and that behavior ranges from not wearing appropriate clothing to a job interview, to interrupting someone too many times in a conversation, to using offensive language at a networking mixer. And I also observe that sometimes when someone acts unprofessionally, others, especially early-career scientists and engineers, think that it is now ok for them to behave in the same manner. For example, if you and I just met at a reception at a conference and we were chatting about our mutual interest in chemical engineering, it would not be professional for me speak negatively about another engineer, converse with my mouth full of food, or pick my nose. All of these actions are unprofessional. And yet, even if I did them, this doesn't mean you should too. Others may be watching you to decide whether they want to engage you, and even I, with my bad etiquette and foul mouth, might grow distasteful of you if you adopt my own negative mannerisms.

But some people, instead of standing strong in the face of inappropriate behavior, are easily swayed to the "dark side." I recall an experiment that a television news program did a few years ago that highlighted this issue: They invited four people to dinner at a restaurant. Another couple also attended, but unbeknownst to the rest of the group, they were there as "plants." Their role was to behave inappropriately at dinner in ways that violate the social etiquette for that culture and the hypothesis was simple: When faced with unprofessional actions, would others fold and follow?

The sextet enjoyed their meal at a round table and the plants took numerous actions throughout the meal that Western culture dictates are inappropriate. For example, they licked their fingers at the table, and they chewed with their mouths open. The pièce de résistance occurred over dessert, when one of the plants held a piece of cake in her hand, took a bite out of it, handed it to the other plant who did the same, who then passed it to the innocent party next to her. Before long the cake had traveled bite by bite around the table. When the experiment was revealed and the parties were asked why they had engaged in such inappropriate behavior, their answer was simply that the other people did it and they didn't want to go against the grain – they succumbed to peer pressure. If this had been a job interview over a meal, I might have noticed you licking your fingers and might have decided not to hire you.

Professional Etiquette in Networking

I have heard some early-career scientists and engineers attest that etiquette doesn't matter. They argue that the major concern in networking is communicating their brand. But how can I possibly think about your brand and the prospect of crafting a mutually-beneficial partnership with you when your unprofessional, inappropriate behavior and etiquette causes a spotlight to shine on the behavior and not the person? That was the case with the candidate above who was interviewing for the development director position. All notions of her expertise went out the window as she disregarded our presence at the interview and instead displayed unsuitable manners for the situation.

The use of proper etiquette demonstrates:

- Your reputation and brand.
- Your professionalism.
- Your strength of character.
- Your respect for those around you.
- Your self-respect.
- Your attention to detail.
- Your values.

> **TIP:** People make subconscious or conscious decisions about you as a professional based on your attention to proper etiquette.

People make subconscious and conscious decisions about you as a professional based on your attention to proper etiquette. This happens in both academia and other ecosystems! So we want to ensure that all they see when they make crucial decisions are your brand, problem-solving abilities, and credentials.

The Tale of the Ball O' Butter[ii]

Years ago, I attended an academic conference when I noticed an anomaly during the luncheon. A gorgeous, well-dressed man had claimed the chair to my right at the table. There are plenty of good-looking academics, but very few of them show up to a scholarly conference impeccably dressed in a three-piece pin-striped suit, matching tie tack and cuff links, and shoes as shiny as mirrors. My reaction upon observing this unusual creature outside his native habitat? This is going to be a mighty fine lunch.

I would like to say that this story has a happy ending and that we united to form scholarly offspring who speak five languages and tell physics jokes without appearing nerdy. But alas, this was not to be in this timeline. As Dr. Suit sat down for lunch, he reached across the table to grab a roll from the bread basket. He buried his entire hairy hand in the vessel until he found the specimen he craved. It was a perfectly round roll. He then proceeded to spread mountains of butter on its entire spherical surface, until the roll itself almost ceased to exist. It had been transformed

... into a Ball o' Butter. Dr. Suit's fingers were smeared with butter and when he appeared satisfied that his masterpiece, the Ball o' Butter, was complete, he then commenced gorging on it like an apple, one huge buttery bite at a time. He shifted said Ball o' Butter between hands, licking his once perfectly manicured fingers as he went. I quickly lost my appetite (for the food and the man).

I often think of this moment – not because I hunger for memories of the grotesque – but because I wonder: Is this how Dr. Suit behaves on a job interview? Or at dinner with his dean? Is this his standard practice when interacting with people he meets for the first time, like me? I would hope not, but something tells me he had no idea that he was demonstrating improper and disrespectful manners, in the process making a lasting negative impression on me.

Professionals in any field often neglect a basic understanding of proper etiquette in interacting with other human beings. We are inclined to argue that our skills, talents, and reputation alone will secure us advancement opportunities. Academics especially opine that any impression they impart from behavior is inconsequential to what super star scholars they are, and it matters not how they hold their fork or eat their bread at a business dinner.

But the truth is being an academic is a profession which means one must behave professionally at all times in the company of the public. As I mentioned above, being professional means demonstrating you are serious about your craft, and having good manners and proper business etiquette for all occasions promotes and amplifies your level of professionalism. When you practice flawless etiquette, your talents are bolstered, allowing attention to be paid to you and not your slimy buttery fingers (which you keep wiping on your pants). Furthermore, in acting as a professional with professional behavioral traits, you are demonstrating a high level of respect for both you and your colleagues.

In Dr. Suit's case, he made some terrible and basic mistakes when he sat down at the lunch. He ruined his chances of communicating his wisdom because all I could concentrate on was his bad manners. Here are some pointers for professional etiquette at meals and in interactions so that you don't become a Dr. Suit:

General rules:

- When dining is involved, the meal is never about the food. When I was in college, every networking function centered on the cuisine. I used to organize my days around going to publicly-promoted mixers where I could get free meals of hors d'oeuvres and other finger foods. I was driven by it and was proud of my accomplishment of being able to go a week without buying a meal. But I am not in college anymore; and now I know that food, whether it is free or not, is not my motivation for attending receptions and other networking events. This doesn't mean you shouldn't eat at a reception – they are supplying the food after all and you can enjoy it. But you should follow some tips to make sure that what you eat doesn't get in the way of your talking, meeting new people, and making networking connections (see tips below).

- Dress the part. Know what is appropriate clothing to wear to an interview, event, mixer or conference, and take it up a notch. An important part of establishing your professionalism and your dedication to your craft is wearing the garments associated with the vocation. However, at conferences and other high-impact networking situations you don't want to appear as if you just came in from the field or lab. You need to appear a little more formal and polished when trying to make a great first impression. This is especially important if you are going to be "on stage" – giving a talk or a poster or leading a board or committee meeting. So where some might consider it appropriate to wear jeans and sneakers to a conference, if you are giving a talk or are looking for a job, especially in the early stages of your career, I would dress up just a little. You don't have to wear a three piece black suit, but a nice pair of unwrinkled chinos combined with a button-downed shirt (tucked in) and a pair of dress shoes is completely suitable for many science and engineering conference settings and even some job interviews.
- Every culture has its own rules – learn these before you travel or interact with someone from that region. How you interact with others, especially those who come from a different culture or region than you, can make or break your next encounter. And since science and engineering is a global enterprise and will only continue to be more so in the future, it is critically important for you to gain an understanding of cultural nuances and norms as they relate to professional interactions before you endeavor to join a team or work with someone from that culture.

I learned this the hard way when I was studying abroad in Cairo. When I first arrived in the Middle East, not yet even 21, I ventured to the souk, or marketplace, and started buying souvenirs. But having not done much research relating to the culture of the region and how business is done there, I went about it in all the wrong ways and as a result I not only paid higher prices for my objects of desire, but probably ended up insulting the shopkeepers in the process. But by the end of my semester stay, I knew exactly what I was doing, having observed people multiple times making transactions and from asking questions (the value continues!) of my classmates. So in December, I recall visiting a merchant and spending time with him at his booth: Chatting with him about the weather, school, family, culture, and the like, enjoying a fresh glass of carrot juice and tea, and then quite some time later, and only then, actually getting down to business and beginning our formal negotiation for the price of the product I wanted to buy. By learning, mastering, and ultimately employing appropriate etiquette for the culture in which I was a visitor, I was able to foster a fair exchange of both product and respect with this salesman.

I learned a lot about operating in unfamiliar cultures from that study abroad experience. For example, I learn in the Middle East to always shake hands with your right hand, as the left one is considered unclean, and to never point the soles of your shoes at someone, as it is considered an insult. I also learned never to assume anything about a person's culture or background without inquiring about it first. I gained this bit of wisdom, also in Egypt, while participating in a traditional Thanksgiving dinner with my new friends from the American University in Cairo. The guests at this affair were from myriad cultures and countries with which I had previously not had any interaction. So I was a little surprised when, while we were having a meal of turkey, gravy, mashed potatoes, and peas, one student ate the entire meal with his hands. Furthermore, he did not make use of his napkin until the very end of the meal, when he performed the following act: When the entire plate was wiped clean of all food particles, he poured his glass of water over his hands above the plate, and then and only then did he wipe his hands on the napkin.

I could have guessed that he was probably from India or Pakistan, where many meals are eaten without utensils. As it turned out, he was from Brunei. And following the meal, I asked him about what he did and he was happy to share the nuances of his culture with me and the other guests. The experience taught me a couple of critical things:

- In a multi-cultural engagement, if you see someone do something that seems out of the ordinary for your culture, you can ask – people are often happy to share their culture with you.
- Don't assume that someone is being unprofessional because they are acting in a certain way that seems to be the opposite of what you are used to. They may just not be familiar with the cultural practices of that region or ecosystem.
- Learn and respect other cultures – just because someone does it differently than you, doesn't make it wrong.
- People respond positively to others who seek to adopt appropriate cultural practices. If you ask about, learn, and then follow the nuances of a certain region, you will find that your networking ROI will immediately improve and you will gain access to the Hidden Platter of Opportunities!

TIP: Networking with like-minded professionals is always a privilege.

And one final thought about interaction with people from other cultures: We all know that every discipline of science and engineering (and even subdisciplines) has its own "cultural" norms and standards. For example, what might be appropriate to wear while giving a talk at an ecology conference might not be appropriate when interviewing for a job at a government lab. So take note of what your own STEM culture dictates is appropriate and how professionalism is defined by that culture. You can use it as a guideline for your own behavior and even kick it up a notch (as I discuss below).

- Watch your alcohol intake. Many networking receptions come with free beer. And for those who enjoy alcohol there's often nothing better than a drink that is qualified with the word "free." But in networking it is extremely important for you to remain clear-headed and focused. So if you do drink, drink sparingly – in other words, much less than you would if this were a social occasion. On the other hand, don't think you have to drink just because they are serving alcohol or it appears as though everyone else is enjoying a glass of vino. There are lots of people who don't drink for numerous reasons relating to culture, religion, diet, personal morals, and propensity towards addiction. I personally never drink so I usually find myself ordering a Coke (of course!) or if I really want to get crazy, a Shirley Temple (which is just Sprite and grenadine).
- And finally, have fun and recognize it is always a privilege to spend time with others who are as passionate about science and engineering as you are. Not everyone on Earth has the honor of being able to talk shop with someone who is as passionate, dedicated, and excited about something as they are. You are part of a lucky handful of people who have this privilege – it is afforded to you as part of your role of being a professional scientist or engineer. So enjoy the time you spend networking with others, honor those around you with appropriate etiquette, and seek to truly appreciate this advantaged opportunity in which you are able to partake.

Some rules for professional etiquette in interactions:

- The first interaction: When you meet someone for the first time, there are five things you should do. Introduce yourself, shake the person's hand, look them in the eyes, smile, and say their name back to them (so they know you are listening and you know that you pronounced their moniker correctly).
- Shaking hands: Shaking hands leaves more of an impression than one realizes. Your handshake should be firm, dry, and quick. The shake should employ two pumps up and down, and then get the heck out of there. Don't linger and don't keep holding their hand like you're mates. Don't use your other hand for the "reach around," in which you grab your colleagues shoulder and shake their entire body. Utilize the whole hand – don't engage a shake with three fingers. Keep yourself

dry by not clasping anything in advance (like a drink or a briefcase), and always use your right hand. If you are waiting to meet someone and are feeling nervous, sit with your palm upwards to keep it dry.

- Maintain eye contact: As you speak with someone, try to maintain eye contact as much as possible. You never want to appear as if you are bored with the conversation and thus have your gaze drift aware from the other party. They should know you have their complete attention. This simple act goes a long way in demonstrating a positive attitude.

- Express enthusiasm for the conversation: There's nothing that's more of a downer when you first meet someone and as you speak and share your brand and experiences, they don't smile, they don't nod their head, they don't express any positive emotion at all. Don't be "that guy." Show your companion that you are interested in what they have to say – this will leave a great impression with them and ensure that the discussion continues fluidly in the future.

- Ask for their business card and offer them yours: More information about the business card is below. The key point here is that you should have a business card and you should not be afraid to ask for someone else's card. You need that information to get in touch with them later. But do be aware that there are cultural differences in the way Americans, for example, exchange business cards as compared to Japanese. In Japan, you present your business card very formally with two hands with the text facing away from you (so the other party can easily read it). When the other party takes it, also with two hands, they proffer a compliment on it, such as "what a beautiful logo" or something of the sort. And there is a hierarchy associated with who presents their card first.

- When your meeting concludes, excuse yourself appropriately and bid them farewell – don't just walk away. Don't be like that astronomer who was chatting with me and suddenly ran away when the opportunity presented itself. Conclude your meeting formally – with a salutation and a handshake. And although you should wait until they are finished with their thought (i.e., don't interrupt them), you don't have to wait until the party is over. If there is an appropriate break in the conversation, it is perfectly fine to say "Thank you very much. I enjoyed speaking with you. Do you mind if I follow up with you next week?" (More information about this is in Chapter 7.)

Some rules for interaction over the phone:

- Prepare in advance: Have your questions written out and whatever research materials about the person or their organization in front of you in case you need to refer to them in the conversation.

- Take notes using a pad and pen, rather than a computer: This only applies if you have a noisy keyboard like I do. You don't want to be distracting your colleague on the other end of the phone with your typing.

- Plan to have it in a quiet place: Just because it is an "informal conversation" doesn't mean you should treat it informally or with a lack of attention to detail. Don't have it in a coffee shop, in line at the grocery store or while you are driving or on a train. If you don't have a quiet place to have the call, either plan to go to the library and ask if you can use one of their study group rooms, which are often available, or reschedule the call altogether. Better to have a quiet discussion where the other party can hear you and you are showing respect than have your sister screaming in the background about when you are planning to walk the dog.
- If you are using a mobile phone, make sure that your device has excellent reception in the place you plan to make the call, and is 100% charged.
- If you plan to use the speaker phone option, do a practice call with your friend so you know that your voice comes across clearly on the other end of the phone. Otherwise use a headset so you have your hands free to jot down ideas.
- If the call drops, simply call the person back and apologize and continue with the discussion.

Rules of video calls and Skype

More and more informational interviews and even job interviews are taking place via video services like Skype and Facetime. Here are some tips to ensure you have a successful experience:

- Know what your webcam sees: Is there a pile of junk behind your head? Is there a poster of Star Trek Captain Janeway that would be both embarrassing and unprofessional to display? Practice with a friend so you know what actually shows up behind you and what you need to move away from the eye of the camera.
- Know how the camera sees you: Since I am very short, if I had a webchat with someone and I didn't arrange my seat or camera in any specific way in advance, all the other party would see would be the top of my head. I have to sit on two dictionaries to ensure that my face is in view of the camera, and you should also determine exactly how you are seen. Similarly, you might have to sit a little closer or farther away.
- Wear professional clothing: You don't have to wear a three piece suit and full makeup and jewelry, but you should look like you consider this meeting to be important and professional. So swap the tee-shirt for a nicely pressed shirt and jacket. Keep the colors solid – no patterns, as

they might cause some distracting optical effects for the viewer. Since no one is looking at your legs, you can keep the pajama bottoms (just don't stand up!). And as for accessories, stay as minimal as possible, because certain jewelry might also create a negative effect and distract the other party.

- Know how you sound: Watch your speaker/headset, and make sure, through a practice session with your pal, that you can both hear and be heard with clarity. If that means investing in a headset, do it.
- Don't look at yourself: Maintain eye contact with the camera, not your gorgeous face. Remember this should feel like an in-person discussion. As such, the other party's desire is to look you in the eyes (as is your goal with them). They don't want to see you looking towards the bottom of your own screen, staring at your amazing face.
- It's ok to take notes: Again I would go for a pad and pen versus a computer. And just as in a job interview, you can look down occasionally to take notes, but your focus should mostly be on the other party's face.

Rules of the table: As you'll be attending many, many networking events that revolve around meals and food, here are some tips as to how to maintain your professionalism and your cool in these sometimes confusing scenarios.

- As soon as you sit down, put your napkin on your lap. Keep it there the entire time you are at the table.
- Wait until everyone is seated before eating, even the rolls.
- When passing items, if you are the person who starts, offer the item to the person on your left and then pass it to your right.
- If you are unsure of which utensil to use, start from the outside in, and for each course use the utensil that is farthest from your plate.
- At a round table, use the "b-d" rule to determine which glass and bread plate is yours. When you sit down at a round table, you are immediately faced with lots of glasses, coffee cups, and bread plates. Which is yours? You can't go wrong with the b-d rule. In your lap, take both your hands and form the "ok" sign with your thumb and pointer finger touching to shape an "o." Keep your other fingers extended straight and together. With both hands in this position, you will see the shape of a "b" on the left hand and a "d" on the right. The "b" stands for bread, which means your bread plate will always be on your left. The "d" means drink which translates to your drinking glasses and cup placed on your right (Figure 4.1).
- If someone screws up and you end up without a plate or glass, ask your server for another one.
- Consume the bread one bite-size piece at a time, and butter said pieces as you go. Don't eat your roll like an apple. The courteous way to dine

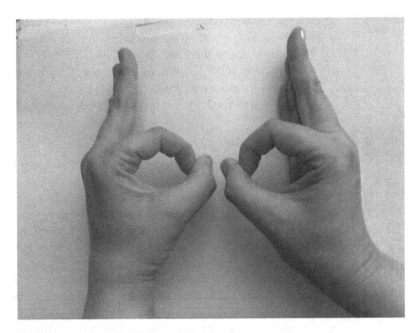

Figure 4.1   The "b-d" Rule.

on bread is to tear off a bite-size piece, butter only that morsel, and pop it in your mouth. Chew, swallow, and repeat. It may take a million years to eat your bread, but at least you will look like a gentleman or lady while doing it.

- If you drop a utensil, ask for another.
- If you encounter a pit or gristle in your mouth, don't put it in your napkin. Simply remove it with your finger and place it on your bread plate (note: some etiquette experts might disagree with me on the removal method – I have been told that however an item enters the mouth, it should exit the same way, thus requiring one to utilize a fork to dispose of the foreign object).
- Cut your food into small, manageable bites, one bite at a time – in other words, don't shove that whole steak or cherry tomato in your mouth and don't cut the meat all at once. Cut what you need for one bite, eat, and repeat.
- Don't lick your fingers.
- Don't reach or grab for something on the table; ask for it to be passed to you and then leave it in front of your plate.
- If you are asked to pass the salt, pass both the salt and pepper, and make sure you don't grasp the shakers from the top.
- Don't pick your teeth or nose, don't apply lipstick or chapstick, and don't blow your nose at the table. If you have to wipe your nose, excuse yourself and do it privately.

- When done with your course, place your fork and knife in the following positions, depending upon where you are:
  - In the States: at the 3 o'clock position.
  - In Europe: at the 6 o'clock position, or put the fork and knife at the 10 and 2 o'clock positions.

  In both cases the knife's sharp edge points into the plate.

Rules of the restaurant:

- When meeting someone in a restaurant, wait for them in the foyer as opposed to the table.
- Don't order anything until everyone has arrived.
- Allow the host to determine what items and price range you should order or adhere to – you can ask them "what do you recommend?" if you are unsure. You should generally order an item that is moderately priced, even if the host recommends the lobster.
- Stick to the basic items to order – beverage, main course, and possibly a salad. Again, let the host determine if you will also order dessert and coffee.
- Choose foods you know and know how to eat. This is not the time to experiment with phal (Indian curry made from the hottest chili peppers on the planet) or jerk chicken, especially if you don't know what the item is or what it could potentially do to your digestive system! This is yet another time when the host can assist you in deciding what to order.
- Maintain the same eating pace as those around you.
- Generally, don't order a sandwich or anything you have to eat with your hands, but if you do, cut the item in half.
- Don't order spaghetti or any other pasta that requires you to suck it into your mouth.
- If you need to use the restroom, wait until the course is over and then excuse yourself. Place your napkin on the table to the side of your plate (not on the seat).
- Don't finish everything on your plate.
- Don't hold your fork like you are in prison.
- Don't season your food before you taste it: During a meal, people make subconscious and conscious mental notes about your problem-solving abilities. If you dump salt and pepper on your squab before testing it, could this mean you will jump into a project without collecting the requisite data?
- Don't use your fingers to eat unless the cuisine demands it.
- Don't ask for a doggy bag.

I was having dinner with one of my students and a CEO a few years ago when I noticed my student was holding his fork like he was in the Big House and was fearful someone would try to swipe it. He treated it like a scoop, and shoveled food into his mouth like it was his last meal. I was embarrassed for him, embarrassed for me, and embarrassed for the business leader, especially since the student was speaking with him about potential job opportunities. I would have hated for this talented, intelligent, and driven student of excellent academic pedigree to miss out on a professional opportunity simply because he did not take the time to employ the most courteous way to interact with someone over a business meal.

The reality is that scholarly strength can get you in the door, but proper etiquette and manners will seal the deal and, ultimately, elevate your academic credentials and lubricate your networking activities. So the next time you have an important function, wear a nice outfit (i.e., appropriate garb), shine your shoes, and make sure you hone your business etiquette skills before you go. You will make an impression that can land you the opportunity you crave. And for goodness sake, under no circumstances, no matter how much you desire it, don't lick your fingers and don't build a Ball o' Butter.

Taking Advantage of Every Opportunity

The key to unlocking hidden opportunities is to take advantage of other opportunities when they present themselves. And although there are an abundance of opportunities, there are very few opportunities which will make themselves known to you that:

- are game-changers, in that they will literally change the course of your career;
- you can pursue right now, in that you have the skills, experience, and abilities to participate, be productive, and succeed in the activity that the opportunity offers at this point in time and space;
- have a wide enough time window, in which you will discover its existence and during which you can pursue it to its full course.

TIP: Most people don't think about opportunities, they don't keep their eyes open for them or they are afraid to pursue them because they might fail.

Many exceptional opportunities appear instantaneously, and often remain accessible for a very small and remarkably fleeting period of time. As I mentioned in Chapter 3, when you see an opportunity seize it now because it may disappear before you know it.

Staying on alert for and taking advantage of opportunities is a science and an art. It can take practice and it is not something that many people naturally know how to do or how to do with finesse. Admittedly, I am a very lucky gal. I learned about seizing opportunities as they came my way

> **TIP:** Opportunity and Networking Node identification starts with things you are interested in.

very early in life thanks to my mother. She both verbally and non-verbally taught me that most opportunities, especially game-changers, flicker; they last only a moment and if you don't have the guts to seize them when they show up, someone else will, and you may lose the chance to pursue it ever again. One of her favorite lines that she would quote to me is from the movie *Auntie Mame*: "Life is a banquet and most poor suckers are starving to death." I took this to heart. It molded my way of thinking. It became part of my mantra and my natural way of living my life.

So I have tried to keep my eyes open for opportunities, whatever they may be, throughout my career and even before. I didn't necessarily look for opportunities that were "career-focused." Instead, my original opportunity management program was geared towards experiences that I thought I would enjoy. But there was one common thread: For every opportunity that I pursued, more opportunities came my way. For example:

- When I was an undergraduate, I saw the chance to apply for a fellowship to study abroad in Egypt and I took it. As a result, not only did I get the Fellowship, but it set me up on a satisfying path to study Arabic and Middle Eastern Studies. I ended up taking four years of the language and gained a significant level of fluency which brought me great joy.
- I saw the chance to serve as a leader in the Society of Physics Students (SPS) and I took it. As a result, I gained experience essentially running a small business and amplified my brand to the Physics Department, so much so that when I returned from my semester abroad and needed a job, I walked into the Communications Director's office and casually asked if she knew of any employment openings. Because she knew my work with SPS, she gave me the job right then and there. And of course that student job led to me being hired out of college as the Director of Communications for the Physics Department, despite the fact that there were other candidates who probably had more experience than I.
- I saw the chance to do research with an internationally-known astrophysicist and I took it. This was more of a training exercise relating to seizing opportunities. I was 17 years old and had arrived on the UA campus fresh from New Jersey. As school was just beginning, I knew I wanted to do some research in astronomy and physics, which at the time was my double major. I went over to the Honors College to see what research opportunities they were promoting and was excited when I saw that Dr. Fang Li-Zhi, a world-renowned astrophysicist who had also just arrived from Princeton University to the UA, had volunteered as a mentor to undergraduates looking to do research.

Buoyed by my mother, I knocked on his door and asked if I could work with him. Not only did he say yes, but he also arranged for me to get a grant from the NASA Space Grant Internship program to pay me. He remained a guide and informal advisor to me even after I changed majors, and ultimately had a huge impact in my career choices. In 2012, when he passed away, I wrote an article about him for APS News, and the very next year, while I was writing this book in fact, I was contacted by a physics professor I had known years ago who invited me to submit my article to be included in a book he was helping to produce about Fang's legacy that would be published in Chinese.

One of the most unusual opportunities that I took advantage of that delivered me almost immeasurable value was also while I was an undergraduate and again looking for a job. I wandered into the pool hall in the basement of the UA Student Union and asked if they were hiring. Turns out they were. And even though I had absolutely no knowledge or understanding of billiards, the boss hired me for the tidy sum of something like $4.15/hour. My job was simple enough: Hand out trays of billiard balls to players, take their money in return, and make sure no one gets into any fights or places their drinks on the pool tables.

> **TIP:** If there is something unique about your background, skills, and experience that distinguishes you from the competition, let people know about it and include it in your brand statement (see below) and even on your resume or CV. It can even be a great conversation starter.

To another person, this might have been just an employment arrangement to earn some pocket money for college and pass the time. But the job came with a very specific perk: I was given two hours of free play every day. So I started shooting pool, even though I had no idea what end of the table to rack the balls, or what the game of 9-ball was about, or even how to hold a cue. I welcomed opportunities to play with the best players in the hall, and after seven months of shooting continuously, especially with players who knew what they were doing and were better than me, I got pretty good. I got so good, in fact, that in the fall I entered my first tournament and took 3rd place. The following winter I represented the university at a regional collegiate pool tournament in Boulder, CO. And a year after that, I took 1st place in the UA pool tournament, playing against the very player who used to wipe the floor with me when I started.

This opportunity to work at the pool hall means everything to me, because the experience taught me a very powerful lesson: If I wanted something and was willing to work hard for it, I could achieve it. I was never enamored by sports, and yet here I was a billiards champion! And I even gained a nickname – they called me Lady Dyn-o-mite.

So if I could be a champion at a game in only a few months of hard work, I knew that I could achieve other exciting milestones in my life if I simply put my mind to it. Now of course the opportunity did a lot more for me:

- I gained business skills and experience working there which helped me land jobs later.
- I gained and sharpened problem-solving skills in business and technical environments (billiards is all about geometry, strategy, and probability after all).
- I gained a novel way to market myself – there are very few women who are billiards champions. And I could tell that when I spoke to others about my victories in this sport, they were impressed. So I leveraged that achievement as a way to distinguish myself from other math majors and later from other professionals. I brought it up in job and informational interviews, not to brag, but to demonstrate that I had a uniqueness and an advantageous background for solving certain types of problems. I placed it on my resumé in a very strategic spot so even a cursory glance at the paper would garner a notice. And people did notice. It served as a conversation starter and a way to break the ice with many, many people years and years after the fact.

So you can imagine if I was able to gain so much from pursuing an opportunity to work for less than the price of a cup of coffee in the basement of the Student Union, what you can do by pursuing opportunities in your own field, discipline, and industry. The possibilities are endless as to where these many opportunities can take you!

But I do recognize that as you get older and more entrenched in your career you become busier and busier and it may not be possible to take opportunities, even those that seem to be absolutely fabulous and will provide exceptional ROI, every time. And that's ok. You can relax and not beat yourself up about an opportunity if for some reason you miss it or cannot take it. (For more on this topic, see my article in *Science Magazine*.[iii])

Failure is the Ultimate Opportunity

Most people don't realize the incredible power that you have when you fail at something, and as a result they don't take advantage of it. When you fail at something, no matter what it is, if you take advantage of learning from it you can gain an invaluable lesson. In some cases, failure is the only way that you might learn strategic information which can catapult you to

> **TIP:** Failure is the ultimate opportunity, especially if you learn from it.

the next level of your career and profession. The key is being able to admit that you failed, and identify why you failed, what factors contributed to the failure, and how you can prevent it from happening again in the future.

Venture capitalists (VC) and other start-up investors love investing in serial entrepreneurs – entrepreneurs who have started multiple companies, some of which may have failed. Why would a savvy investor park his money with a failure? Because he recognizes that someone who has failed in a circumstance, if they are smart and aware, will ensure they don't fail like that again. The entrepreneur has failed on someone else's dime and time, and the VC is more than happy to invest in the entrepreneur with the knowledge that he has learned from his mistakes.

When I was in college, I failed. Yes it's true! I had a desire to be a resident assistant (RA) in the dorms. And at the time, the process to become an RA was rather lengthy: I had to prepare an application, obtain two letters of recommendation, take a "leadership" course where my actions would be observed and graded, and participate in an interview. It was a nine-month extravaganza – I could have created a human being in the time it took me to try and win the job.

Well I was confident and sure all the way through, to the point of being cocky. I was really certain I had this one in the bag. But when the announcements came in the late spring, not only did I not get the RA job, but I didn't even get an alternate position. Basically the university was saying to me that if all students at the UA had left the campus, transferring to another university altogether and there was literally only empty dorms on site, my services would still not desired to oversee an abandoned shack.

Naturally I was crushed and angry. Why the heck did I waste my time on this annoying enterprise when I could have been doing something more worthwhile – learning Dungeons and Dragons, knitting, or any-thing else? I carried that anger with me for a number of years, blaming it all on the system and everything and everyone else – except myself.

But I finally did some retrospection and realized this failure was my fault for two distinct reasons. First of all, in the class, where they were observing my leadership abilities, they were also watching my team-building skills and how well I play nice with others. Being a Type A personality, I volunteered to lead every team, rather than taking turns with others. My failure was not to have better understanding of what the decision-makers were looking for and what the culture of the RA environment is like. Had I done sufficient research, had I done my own observing of RAs within the dorm in which I dwelled and watched how they interacted with their colleagues, I may have gotten a clue that they don't want someone always wanting to lead.

When I recognized this, and how my own inaction, and more to the point my own neglect in conducting sufficient research on the ecosystem's requirements, factored in to me losing this opportunity, it was somewhat revelational. I was able to realize the strategic importance of fully comprehending an organization's culture and norms, and the value in doing research ahead of time to decipher what the organization itself finds valuable.

The bottom line – don't be afraid of failure. Embrace it. If you fail, ask why this happened. Seek to learn from your mistakes. Seek to find alternative solutions to the original problem, even if for the sake of an exercise. Ultimately, failure can help you succeed. And finally, be open to the idea that failure to do X may lead you on a different path where you end up doing Y. That same year I applied for the RAship, I also applied and won the travel fellowship to go to Egypt. In the end I am glad I did not win the RA job.

Self-promotion – The Right Way

As I mentioned above, there is nothing wrong with self-promotion, if you do it appropriately. And most likely you are already promoting yourself – many of your outputs that naturally are associated with being a technical professional are self-promotional in nature. For example:

- If you have written a published paper (or aspire to do so).
- If you have given a talk in "public," that is to at least one other person.
- If you have applied for a job.
- If you have served on a committee.
- If you have introduced yourself to someone.
- If you have done anything that demonstrates to members of the public, and more specifically to your community, your brand, expertise, excellence, skills, talents, and so on, you are already doing self-promotion.

TIP: Networking requires that you talk (not exclusively) about yourself. So get used to and comfortable with this idea.

Networking in and of itself is a self-promotional activity – you are introducing yourself and your brand to someone new, discussing ways in which you can exchange value, and looking to solve problems utilizing your mutual talents. To do this, you have to talk about yourself.

As you promote yourself, you can begin to gain awareness of both hidden and advertised opportunities. So take note that there is nothing wrong with promoting yourself and that it is necessary for you (and STEM scholarship) to advance.

What are the major self-promotional activities you should pursue?

Speaking

Giving a public talk is a significant element of establishing your brand and attitude and enhancing your reputation. The reason is simple – when someone gives a talk, the audience perceives that person to be an expert and a thought leader in the field. If they weren't, why on Earth would they be speaking right now? The perception (and truth) that you are a leader positions you for significant networking ROI and to take advantage of hidden opportunities. And since STEM naturally requires one to be adept at public communications, the more practice you have at public speaking, the better you will become, and the further you can take your brand and advance your reputation. I can't tell you how many times I have given a talk and someone walks up to me afterwards and asks for a meeting because they want to engage me somehow. They envision that there is a manner in which we could work together and they are hungry to pursue it. This happens all the time.

Offer to speak at conferences, workshops, and meetings no matter how many people are attending. Offer to speak at campus venues such as journal clubs, student clubs, departmental meetings, department recruitment and outreach events, and of course in classes. Offer to speak in the community for volunteer associations associated with your industry and in meetings or at venues where you would like to infiltrate a specific network. And if you are uncomfortable speaking, seek out low-stake environments first such as meetings within your own research group, where you can be comfortable practicing and gain valuable feedback on your delivery. Another terrific resource for up-and-coming speakers is Toastmasters. This is an international organization that frequently has multiple chapters in a municipality and even on college campuses. Its prime directive is to aid professionals to become better, more confident public speakers and thus they offer many opportunities for members to volunteer to give formal and informal, prepared and improvised speeches. All of these venues in which you can speak are networking nodes, which will provide the best bang for your networking buck.

I recently wrote an article where I profiled a scientist who had a desire to work in her home country after she finished her graduate studies and post-doc abroad. She reached out to her undergraduate mentors and offered to give colloquia at her alma mater anytime she returned home. Soon she was branching out to other academic institutions to give talks, which led to invitations to speak, as her reputation as a stellar leader in her field expanded across the nation. By the time she finished her postdoc she had a job waiting for her back at home, and it was in no small part due to her high-impact networking which was aided by her many presentations.[iv]

Writing

It is not enough to write peer-reviewed papers in journals. We want to expand your brand and amplify your reputation to new publics and new networks. And in addition, we want to gain new skills in communicating our scientific and engineering expertise to non-expert publics in a way that elucidates the thesis and makes them want to learn more and even advocate on behalf of the subject and the profession. So think beyond the journals and consider writing any of the following:

> TIP: When you write an article, you create your own networking node.

- Articles for your institution's newspaper, newsletter or website. If you are in academia or a government agency or laboratory, there is undoubtedly a publication(s) that the institution's media relations or even the student government supports. So contact the editor and ask if you can write an article for their publication. They will probably be thrilled to hear from you, as it is often hard to rally writers for institutional magazines. And even the student newspaper is eager to print voices other than those of its usual reporters. Furthermore, if you can tie your research to something current, or something newsy, then it is considered to be even more valuable to the editor and it will probably get more page views too.
- Articles for your local newspapers (like the regional business paper), including opinion pieces or guest columns relating your science and engineering background to local issues, such as business, education, the environment, and politics. Just recently I saw an ecology professor's article in the *Los Angeles Times* about how Hollywood gets animals and bugs all wrong in the movies. What a great topic! The writer was able to convey her passion for zoology and biodiversity and connect it to something an interested public could understand and relate to: the movies. I noticed the article syndicated in my local paper and liked it so much I tweeted it. That tweet was retweeted by her university, and on it went!
- Articles for the membership publications of the associations to which you belong. These can be about anything that you could conceive a reader would find valuable, such as tactics associated with the act of being a scientist or engineer in that field; a tale of an experience you had while mentoring youngsters; tips for navigating the annual conference for first-time participants; or opinion pieces that tie your discipline to something in the news. For example, after Hurricane Katrina hit New Orleans in 2005, I authored a guest column in *PR Tactics*, the monthly newspaper of the Public Relations Society of America. My article was entitled "Public Relations is Human Relations" and discussed the basic

point that in a crisis we must remember to treat our customers and publics like the human beings they are. Given the timeliness of the piece and its relevance to the audience of PR professionals, the publication's editor was happy to receive and publish it. That one article put my name in front of approximately 20 000 members of PRSA in a special issue devoted to crisis communications during the hurricane. (In fact, the editor liked my article so much that its title was plastered on the front page of the magazine.)

- Articles for popular science magazines, websites, and blogs. Many popular science publications welcome content written by scientists and engineers themselves, as long as it is well-written and contains the appropriate level of detail for the audience. The key here is recognizing that the readership consists of people who are science-enthusiasts, but who are not necessarily science-educated. So you have to watch your use of jargon and acronyms carefully. Don't talk down to the audience; rather think of it as a way to inspire and excite a new section of the population about your area of expertise. So use language that helps them understand why you are passionate about it and why it should be important to them as well. *Scientific American* is an example of a publication that relies on scientists and engineers for its main features, but there are many others as well that would welcome your pitch and submissions.
- Articles for community and hobby-based newsletters. Even an article in a neighborhood association newsletter, or a website devoted to people's love of dogs, can open doors to hidden opportunities and to new diverse networks.
- Your own blog – see Chapter 8.

Join organizations, and volunteer for leadership and committee assignments

The goal here is to not only give you new skills, but to shine a light on your existing brand for those in your industry to see. When you serve on a committee, people get a chance to see how you work, what qualities you have, and how your attitude manifests itself. So it gives them a glimmer of what you might be like as a long-term partner, either in an employment scenario or in another career context, and thus it is a very valuable self-promotion opportunity. It also gives you the chance to do some high-impact networking with the other committee members. I have served on tons of committees throughout my career, within my institution, in my region, in my industry, and in combinations of all of those. And honestly, just like many other opportunities, very few people offer to serve, quite frankly, because they either think they don't have the time, or don't want to put in the time or don't see a benefit to them. So when you volunteer to be on a committee it immediately shows people that you have a great attitude towards hard work and the organization; that you are a self-starter and a go-getter; that you are interested in contributing to your community, and so on.

Just recently I joined the membership committee of the one of the professional associations to which I belong. In our first meeting, which took place at the annual conference, I had the chance to meet with other members and get to know them more intimately than I would have if I perhaps just saw them in the hallway. As a result of attending just one meeting, I gained strategic information about hidden career opportunities that I could pursue and solidified my relationship with several members such that if I call them in the future or request a meeting, they will probably agree.

And then, as you work your way through the committee, you can at some point volunteer or run for a leadership position. As the leader of a board, you are seen by your community as a leader. As such, this magnifies your brand and reputation to the decision-making publics and grants you other opportunities, many of which are hidden.

This very scenario happened to me while I was working at the UA. I volunteered to serve on the College of Science Staff Advisory Council. This led to me becoming the chair of this committee. And this ultimately led to the Dean of the College of Science noticing my contributions and calling me in to his office one day to invite me to apply for a new position he was creating. Now as it turns out, I interviewed for the job and didn't get it for various reasons. But a year later, the Dean left the university and a new Dean was appointed who already knew me by my work and reputation. He transitioned the person who was in the job I had applied for into a different department and essentially handed me the position I had pined for the year before. Don't you just love it when you can see the ROI of self-promotion, networking, and following an opportunity-chain?

Volunteer at conferences

See Chapter 6.

Apply for awards

Awards are an important element of being a professional. Winning awards and honors signals to others in your community that you are a success and that your brand is one of true excellence. Awards can open doors to new career opportunities, especially those that are hidden, and can greatly help you expand your networks.

But the majority of professionals generally do not apply for awards, big and small. I am certain that the number one reason that people across all fields, industries, and sectors do not apply for awards is because they fear that they will not win. "There must be someone better than me," they argue and as a result they don't apply. And yet the reality is this: Most of the population feels this same way! Everyone, even and especially those who are most successful at their profession, has a tiny bit of trepidation when it comes to pursuing the next level of advancement, reaching out and introducing themselves to a stranger, and applying for an award. They think they aren't good enough. I have heard geniuses even admit they aren't better than the rest. This is often referred to as "imposter syndrome," whereby people know they are successful in their profession and field and yet feel like an imposter and that they don't deserve the accolades.

But that is hogwash! Poppycock! If you are reading this book you must be a scientist, engineer, or someone who is passionate about these disciplines. It also means that you want to improve your skill set and expand the field of opportunities you can pursue to practice your passion. That immediately indicates that you have the skills, the experience, and the drive to be successful and you most likely have already been touched by success many times throughout your career, and perhaps didn't even recognize it.

> **TIP:** You have to be a passionate action advocate for your own career. Find the opportunities you need or want and do what it takes to get them. Don't wait for them to be passively handed to you.

So you are "good enough" and "smart enough," as an old joke on Saturday Night Live attests. So let's take your career and networking opportunities one step further by applying for awards. We start by recognizing that we have to be passionate advocates of our own careers, which means we can't sit idly by and wait for someone to nominate us an award or think of us for a fellowship. We have to research the available awards, decide which are right for us, and apply, apply, apply.

What's the importance of an award?

- It validates your excellence and leadership in the minds of your community.
- It serves as a credential that confirms that the community thinks your contributions are superior.
- It serves as a legitimate means to promote your success and recent achievements.
- It elevates and amplifies your brand, attitude, and reputation to key publics.
- It opens access to the hidden platter of opportunities.
- It naturally creates and fosters networking alliances with new populations.
- It can lead to winning other awards.
- It helps you hone your communications skills.

And here's another tip – even if you don't win the award, you have gained some very specific self-promotion and networking returns. First of all, at least one person (and probably a committee) has read your application packet and learned about your career victories, so in a very direct way you are promoting yourself to this team.

Another misunderstanding about awards mirrors what novices think about networking: That it is a one-way street of benefits that are only acquired by the award winner. In actuality, the award process is just like networking in that its intention is to create a win-win partnership with the community and specifically with the applicants and winners of the honor. In fact, organizations that establish awards programs reap a great deal of their own public relations rewards, including:

> **TIP:** For an organization bestowing an honor, the award process is just like networking in that its intention is to create a win-win partnership with their community and specifically with the applicants and winners of the honor.

- Promotion of their mission, programs, and projects.
- Elevation of their brands in the minds of the community.
- Demonstration of their dedication and commitment to serving the community.
- Creation of relationships with new advocates and solidification of current partnerships with people who support the mission of the entity.

Finally, and most importantly for you as you look to expand your networks, an awards program helps the organization learn about stars and rising stars in the field who could potentially be assets to their organization. For example, I have applied for awards and as a result the decision committees (and by extension the sponsoring association) got to know my brand and saw that there could be value in partnering with me in other ways. Because of this I was invited to serve on awards committees, write guest articles, and participate in other programs, all of which opened up new networking vistas for me.

> **TIP:** Most people are afraid they won't win an award and therefore don't apply. Therefore, by simply applying you are putting yourself in an advantageous position. As the Lottery motto goes, "You can't win if you don't play."

Awards to apply for:

- International and national awards directly related to science and engineering research innovations.

- Teaching awards.
- Outreach awards.
- Awards offered by your professional society.
- Awards offered by your current institution.
- Local business awards, such as the 40 Under 40 Business Leaders in a region (most local business newspapers have lists like this).
- Web-based or social media awards.

Types of awards:

- Grants – think large, small, and tiny – even small pockets of money hidden within the bowels of your institution are fair game to pursue.
- Fellowships.
- Competitions and contests – which require you to not just submit an application but an example of your body of work.
- Awards in name only.

And don't think that every award you apply for has to be a "serious" honor like the Nobel Prize. For a number of years, *Science Magazine* has sponsored "Dance Your Ph.D," "an international competition to see which scientists can best explain their graduate work through interpretive dance." Granted, if you win this honor, it alone probably won't secure you a job offer from Cambridge University. However, it will give you notoriety, a distinctiveness, and a bit of a niche in which to market yourself. Furthermore, given that the contest has a specific communications component, if you win it demonstrates your unique communications acuity and creativity. It can certainly act as an ice breaker and a conversation starter. It can help you win a meeting with a potential employer who can offer you a formal interview. And it can certainly give you access to hidden opportunities, because it exposes your new talents to new potential collaborators.

Tips about awards:

- Consider the potential ROI before you apply and weigh any risks associated with applying, winning or not winning the honor. This will give you the chance to see how much time you should devote to this enterprise and whether it is indeed worth your time and energy. But remember the bigger the risk, the bigger the reward, so often the awards that involve the most complicated requirements or take the longest to complete the applications, are the ones that deliver more on their promise of promotional and networking opportunities. But ultimately you have to decide for yourself whether the time you put into applying for an honor could be better spent in the lab or sending out cold emails for traditional networking activities.
- Most often you can nominate yourself, and don't hesitate to do so: In the United States, there is usually nothing wrong with nominating

yourself for an award, especially if the award criteria specifically state self-nominations are welcome. But if you want to apply for the honor and it requires someone else to formally nominate you, find out what the nominator needs to do and ask your trusted mentor if they would consider nominating you. You can then offer to do all the basic work for the award (see my note below about letters of recommendation). Remember, you must be your own advocate, as your career is at stake.

- Connect with the awards' chair or administrator in advance: You will get the inside scoop on how to tailor your award application and what to highlight and even leave out in your packet. It also gives you the chance to make the acquaintance of the professional in charge, who will then keep an eye out for your application. They may even help shepherd you through the process or provide you with other opportunities. This happened to me recently. I had applied for an award two years in a row and had not received it, but after each rejection, I contacted the organizer to determine how I could improve my application for the following year. By doing so, she got to know me and she could see that I could be an asset to the organization. So by the time I applied the third time around, she wanted me to win. But there was a problem – that year there were so many top notch applicants in the program I was applying for, there wasn't a spot available for me. So she called me and said "we really want you for this award, and I know you had applied for X program, but would you consider participating in the Y program instead?" Both programs were equally exciting, I just happened to be more interested in the subject matter of one rather than the other. But the prospect of being named an award winner in Y was way too good to pass up. And I realized that by going out of my comfort zone and choosing Y, I might just gain a lot more than I would have if I had only been able to get the X award. So I happily and gratefully accepted. The Y award program was so fantastic and introduced me to whole new concepts of my profession that I didn't know anything about. It gave me access to professionals in new networks and shined the way towards other hidden career-changing opportunities that I could not have predicted. It even led to very specific job leads!

- If you are asked to apply, do so! If you get to speaking with the award program manager, and he/she invites you to apply, they are probably doing so for one of two reasons: There might not be enough people applying for the award, or they know you are a superior candidate and probably will win. Now of course they might just be trying to get as many applications in as possible. But quantity of applications is not necessarily better than quality of applications, insofar as the award sponsor is concerned. In other words, they usually would rather have 5 superior applicants versus 38 subpar applicants. The former solves their problem and makes them look good in front of their boss, whereas the latter calls into questions whether the awards program should be sponsored at all.

- If you need a letter of recommendation or nomination, offer to write a first draft: Your mentors are busy people and may not have the time to spend researching the award and your background and then writing a customized letter that will clarify everything the committee needs to know to make a decision. So you do a great service to your supporter to offer to write the first draft. In this way, you can include the relevant information about your experience, expertise, and successes that they might not even be aware of, but are crucial to the awards committee. This is generally an accepted practice in academia and beyond, especially in the United States, so don't be shy about it. But if your mentor prefers to write the letter themselves, provide them with as much information as possible, including your CV, the award name, sponsoring organization and criteria, and any other information you have gleaned from your own research of the award or interaction with the award administrator. You can request your mentor mentions specific details of your experience that are relevant to the award. And don't forget to highlight when the letter is due, what form it should take (pdf, email, or typed letter), to whom it should be addressed, and where it should actually be sent (to you or directly to the awards committee).

- If at first you don't succeed, try, try, try, and try again: There have been many, many awards that I have won in my career(s) that not only did I not get on the first shot, but didn't win on the second or third shot either. For that matter, there are certain fellowships that I have applied for at least four times and still haven't received. But that won't stop me from applying next year. Each time I have sharpened my abilities in communications and self-marketing, added new allies, and expanded my networks. And award committees recognize tenacity, often favorably – I have received letters from award administrators apologizing for not being able to award me the honor, but encouraging me to apply next year or to look at other awards that their organization sponsors.

- Follow up with the awards committee: If you win, send them a thank you email. And if you don't win, send them a thank you email and ask for feedback about your application. Often (depending on the nature of the award) the administrator will give you an honest evaluation of your application and suggest ways that you might improve in the future. Here is what I would say if I didn't win an award:

Dear Sydney,

Thank you very much for your consideration of my application. Although I am disappointed I did not win the honor this year, I completely understand that the competition was very strong. I intend to apply next year and would appreciate any feedback you could provide that will help me enhance my application in the future.

   Thank you again for your support of this important endeavor. I appreciate your thorough perusal of my application and look forward to perhaps working with you next year.

Best Regards,
Alaina

- Send thank you notes to your supporters: If someone took the time to write a letter of recommendation or endorse your application in some way, send them a thank you note, preferably handwritten. And you can do this whether you win the award or not. You certainly need to express gratitude and you also want to keep them abreast of your progress in this particular endeavor.
- If you are able to find out who wins, send them a note of congratulations: You don't have to mention that you didn't win; simply express congratulations on their success. Very few people who win awards get emails from people they don't know congratulating them (or for that matter even people who do know them!), and they will appreciate that you took the time to do so. You can now follow up and leverage this opportunity to add them to your network.

As an undergraduate, I used to walk by posters plastered across the campus announcing the US Department of

TIP: If someone else can do it, you can do it too!

Defense National Security Education Program, a prestigious fellowship for students eager to study abroad. The program, now called the Boren Fellowships, "support[s] study and research in areas of the world that are critical to U.S. interests, including Africa, Asia, Central & Eastern Europe, Eurasia, Latin America, and the Middle East." I had been fascinated by Egyptology since before I could identify Egypt on a map, so the prospect of getting an all-expense-paid study abroad experience to Egypt enamored me. But at the same time, I actually said to myself "there's no way I could ever get this award. It's too prestigious. It's a national award and there's way too many people applying for it who are probably better than me." So I didn't apply for it. And then the spring of my sophomore year I was sitting in English class with my friend when she announced that she had won this fellowship and was going to be spending a year in China, her dream country and culture. My immediate response was delight for her and outrage at myself. How could I ever have been so narrow-minded to assume that I wouldn't make the cut? I swore to myself: If she can do it, then I can do it. And I did – I applied for the fellowship six months later, pulling an all-nighter and even skipping class to finish my two requisite essays. And six months after that I got the letter signed by the US Defense Secretary himself congratulating me on my selection. I was going to Egypt.

As I mentioned above, that one award, that one opportunity, opened the doors to many more career opportunities and networks that have made my life all the more enriching. It gave me a credential which demonstrates to others that I can work on multinational, multicultural teams. It bolstered my applications for other honors and helped me win scholarships. It gave me a unique niche in which to market myself, because very few undergraduates study abroad in the Middle East as compared to the UK, France, and Italy. It gave me access to professors, scholars, and other experts in Middle Eastern Studies who provided me with knowledge that helped me make informed career choices. The list of benefits I received from that one fellowship goes on and on, and continues to this day.

Promote your honor

When you win an award and you appropriately promote the achievement, it can serve as a conversation starter and a way to connect with new contacts or reconnect with current contacts. For example, in 2012, when I won a journalism fellowship to attend the Nobel Laureates Meeting, I used that piece of information as a way to follow up with contacts with whom I hadn't spoken recently. I wrote them emails with "great news" in the subject line and emailed them about the wonderful bulletin that I had received this award and felt humbled to be able to attend such a prestigious event. Almost everyone I wrote to emailed me back offering congratulations and asked to be told when my articles from the fellowship would be published so they could read them.

Promote your honors and awards by listing them in:

- your industry/trade publication,
- the "member news" section of your professional society,
- your alumni magazine(s),
- your local business section of the local business newspaper,
- on your LinkedIn profile, Facebook page, and Twitter page,
- as an update on LinkedIn, Facebook, and Twitter,
- on your resumé/CV,

     o   if it's something really big – like a Nobel Prize – consider putting it on your business card.

- Attend the awards ceremony (If there is one): These give you a great chance to network and to meet leaders in the field, whether you win the honor or not.
- And finally pay it forward! If you know of an award for which you don't plan on applying, or think would be perfect for a colleague, let them know about it. Even if you are both competing it is still an appreciated sentiment to share news about this opportunity. You can even offer to write a letter of recommendation on their behalf. They'll be positively pleased, you will have solidified your partnership, and when they win, they will remember that it was you who helped bring them this good fortune. And when they win, help them by promoting it to your own networks and congratulating them publicly, such as on LinkedIn.

Social media

See chapter 8.

Have a business card

No matter where you go, it is always helpful to have a business card that you can hand to someone. It condenses your contact information, it demonstrates your professionalism, and it shows someone you are very serious about your work. Many scientists and engineers, especially those in academia, don't think to print business cards or to have them handy at a conference. That's usually because they don't think they need them. But if you are conducting strategic networking, and speaking with as many people as you can at as many events and other opportunities, you absolutely must have business cards available to exchange. After all, you want to be able to give them something they can hold on to, that is a physical manifestation of your brand. And since you can't hand out 36 000 copies of your dissertation every year (nor would you want to, and nor would other parties want to take them), the business card is the simplest means to communicate your contact information and your brand to people you meet. Now of course, technology is changing very rapidly and is allowing for new means of getting another party's contact information. I have seen people whip out their smartphones at mixers and type in my email address right into their digital phone book. But this currently is the exception to the rule. For the most part, exchanging business cards is still the norm.

So what do you print on your business card? They don't have to be fancy. They can simply have the following information on them (Figure 4.2):

- Your name.
- Your "title": by this I don't necessarily mean your job title. You can certainly use your job title if you want, but what do you include if you are unemployed? Or perhaps you are looking for a job outside of your institution and even outside your discipline. Your title can simply be your brand statement – it describes who you are or what you do. For example:
  - Candidate, PhD, Wildlife Biology
  - Molecular Biochemist
  - Aquatic Ecologist
  - Physicist, Science Writer, Speaker.
- Your expertise: You can list subdisciplines and subjects (like elementary particle theory or fish biology), special skills (like C++, Java, or Python, or Next Generation Sequencing or electron microscopy), or even the names of tasks that you do (like blogger, teacher, author, or comedian).
- Your email address: keep it clean folks! No sexyscientist@yahoo.com. Have a professional email address – in fact here's a tip – have an email

Figure 4.2   My Business Card.

address that is separate from your work address (which is very important if you are looking for a job while still employed).
- Your phone number.
- Your LinkedIn profile URL: See Chapter 8 for more information about this.

Some thoughts about business cards:

- If you are with an institution, often you can get them made for free. As a professor and possibly even as a postdoc, your department will often pay for you to have business cards made with the organization's logo.
- Don't put your picture on it: Generally in the United States, a picture on a business card is confined to those who are real estate or insurance agents. A real estate broker once told me why this is: The idea is that once you see my picture on my business card a few times when you glance at how gorgeous I am, your brain will start to think that we are friends, and you will be more likely to trust me. That's all well and good for real estate, but for other professions, your photograph is not appropriate. (Note that for a LinkedIn profile, a picture *is* totally suitable *and* a standard requirement, and in fact boosts your visibility on the site, which I discuss in Chapter 8.)
- You don't need a logo. You can put a picture of something related to your work, such as a piece of equipment or a scene from a microscopic slide. If you are associated with an institution you can use their logo (although it is always good to check about permissions for doing so before you have them printed yourself), but really there is nothing wrong with a business card that has only simple, nicely formatted text and no image.
- You don't have to use fancy paper or specialty printing. Yes, a silver encrusted, embossed outline of a microscopic image of a telomere is totally cool, but is it worth the extra amount of money you would pay to put it on the card? I would argue no. And as far as the paper is concerned, make it a decent weight (in that it doesn't easily fold) but it also doesn't have to be a high-end weight of paper, equivalent to super thick card stock.
- You can even print them yourself. The technology and the paper options for printing business cards at home have jumped by leaps and bounds in just a few years. In ancient times, if I printed my own cards at home, the paper I purchased was super thin and had micro perforations that you could actually feel with your fingertips. They looked cheap and the message they delivered was one of being unprofessional. But nowadays, you can buy business card paper that is thick to the touch and separates easily from the parent sheet without any perforations, so it is perfectly suited to your needs.

Chapter Takeaways

- Once you have identified and clarified your brand, you need to engage in reputation amplification activities to ensure others know what value you can bring to them. This is the cornerstone of smart networking and gaining access to the Hidden Platter of Opportunities.
- Always ask questions, and not just of people in your field – you never know what kind of information and inspiration you are going to get, and there's no such thing as a stupid question.
- The mentor–protégé relationships you have are an invaluable element of career advancement, and each one provides precious opportunities for networking, identifying relevant networking nodes, and accessing the Hidden Platter of Opportunities.
- Demonstrating your professionalism in all ways, in communiqués, in how you speak, what you say and through appropriate professional etiquette, speaks volumes about your brand and your seriousness about your craft.
- Take advantage of as many opportunities as you can to learn more about your field and to advance both yourself and your colleagues, as each opportunity is a chance to engage in networking and each opportunity leads to another.
- If an opportunity doesn't exist, create it yourself.
- Self-promotion and networking are inherently linked; learn the appropriate avenues and methods to promote your brand to decision-makers.

Notes

i.   Escape from the Zombie Boss or Colleague, AGU, http://sites.agu. org/careers/2013/07/22/escape-from-the-zombie-boss-or-colleague/.

ii. Taken from "An Etiquette Lesson," InsideHigherEd.com, August 268, 2006 (link no longer available).

iii. For more on this topic, see *Science Magazine*: "Opportunity Knocks: But Which Door Should You Open?" http://sciencecareers. sciencemag.org/career_magazine/previous_issues/articles/ 2013_03_08/science.opms.r1300129.

iv. From "Postdoc Advancement: marketing your value," Sciencecareers.org.

5 Developing Your Networking Strategy

We have explored various ways in which we can accentuate our brand to diverse publics, who could potentially be decision-makers and offer us access to the Hidden Platter of Opportunities. In this chapter we develop a comprehensive networking strategy and plan to connect with these professionals, engage them in collaboration, and ensure an exchange of value between parties.

Identifying your Goals for Networking

As is the case with anything in your career, it is always critical to begin by establishing goals. An all-inclusive networking strategy is also dependent on goals, which can be organized into three categories:

- Opportunity-centered goals: What you want to achieve from participating in a particular opportunity (such as attending a conference, publishing a paper, serving on a committee, or having coffee with someone).
- Short-term goals: What you want to achieve, at any time from the next few months to a few years.
- Long-term goals: What you want to achieve in the next five or ten years or more.

These objectives will shift in terms of their priority through time, depending on where you are in your career path and what you need to achieve at any given moment. For example, your networking activities may be influenced by any of the following:

- Need a job.
- Need/want advice – about an opportunity, job, organization, industry, region, culture.

Networking for Nerds: Find, Access and Land Hidden Game-Changing Career Opportunities Everywhere, First Edition. Alaina G. Levine.
© 2015 John Wiley & Sons, Inc. Published 2015 by John Wiley & Sons, Inc.

- Want to know about the industry, company, community.
- Want to increase your skills.
- Want/need a collaboration.
- Want/need a grant, financial support.
- Want/need inspiration, new ideas.
- Want to meet like-minded individuals, groups.

Your networking goals will be closely tied to your career goals and will also be contingent upon your desired timeline to achieve these goals. For example, if you know you will be finishing up a postdoc appointment in the next six months and need a job immediately after, your networking goals and actions will be focused on you getting a job in that period of time. If you are a mid-career professional and are interested in moving to a new city or starting a career in a new field in the next year, then your networking goals and activities will be centered on this objective.

On the other hand, just because you land your dream job at age 30 (or however old you are), it doesn't mean you should stop networking. As stated above, networking is essential to continually move your career and scholarship forward. You can't expect to be innovative in your dream job if you don't have a constant influx of new sources of ideas and inspiration. Imagine if all you knew and focused on was what was in your lab or office. Your productivity could potentially become very stale and even stymied. Therefore, as an employee, in any arena, it must be your priority to ensure that you are seen as a valuable (if not invaluable) contributor to the organization at all times. Constant networking, to solidify new partnerships which can aid you in solving problems in novel ways, will benefit you at every stage of your career. And chances are it will also help you decide what opportunities you can and should pursue in the next phase of your career.

As you contemplate your short-term and long-term career goals, you can create specific networking strategies that adhere to the timelines and deadlines you have established for yourself. But please note, while timelines and deadlines are important for any goal-setting agenda, don't beat yourself up if you set a mini-goal to send two cold emails a week and in the first two weeks you send exactly zero. Simply reexamine the reality of your time management system and adjust your networking schedule and goals accordingly.

> **TIP:** Develop a networking strategy that fits your goals and your available time allowance. Make it flexible so you can adjust it as needed, depending on other time commitments and changes in your goals.

It is important for you to see milestones – that is for you to be able to see evidence of your networking activity – in order to bolster your confidence. As such, create a plan that you can actually stick to. For example, if I know that I can only devote 30 minutes a day to researching potential collaborators and sending out cold emails, then that is what I will base my objectives and related activities on. And if next week I have to spend extra time doing something else and don't get a chance to complete this week's assignment, then so be it – there is always tomorrow. But the key thing here is to write down your goals so you can visualize the route you need to take to achieve them and make note of your achievements so you can see you are working towards your overarching goals.

Your networking plan should also naturally allow for a great deal of flexibility and be nimble enough to change as new opportunities arise. You may set a goal of sending out two emails a week, but if this week you don't send any, because instead you volunteered to drive the colloquium speaker to the airport, that's fine! In fact, it's more than fine – you should pat yourself on the back that you took advantage of a hidden opportunity to have private face time with this professional. Your conversation in the car may make you aware of other opportunities and networks that you can pursue, and you should modify your networking plan to allow yourself the chance to do so.

Your networking strategy should be built on the following framework:

- Identify what career goal(s) you want to achieve.
- Analyze and pinpoint what network(s) and networking nodes could help you achieve this career goal.
- Establish networking goals to access the network(s).
- Find contacts in the network(s).
- Initiate contact requesting "informal conversations" or "informational interviews."
- Conduct the informational interviews.
- Send thank you notes or emails.
- Follow up.
- Organize your networks.
- Continue to follow up and look for opportunities to exchange value over time.

Let's analyze each element of the framework.

Identifying what career goal(s) you want to achieve

Your career goals will be very much contingent upon your self-assessments that you completed in Chapters 2 and 3. As you analyze the patterns that emerged in your Skill Inventory Matrix and your SWOT Analysis, you can begin to think about career paths for yourself. If you are unsure of what careers might be right for you, let your networking guide you. Up until I was in my senior year in college, I was certain I was going to go to graduate school in mathematics and/or anthropology and stay in academia. But as graduation loomed, I came to the firm realization that I was not interested in a life consumed by mathematical research, nor was

I thrilled about pursuing the profession of a professor. But as I mentioned above, I had no idea what else I could do with my education and training. The reason I was ultimately able to carve a path in science communications, which led to every other career choice I have made, was because I was enthralled by and attracted to science outreach, writing, public relations, and marketing, even if I didn't know that there was a job with the title of "science communications specialist." I followed my interests towards opportunities that let me utilize "science communications" skills which led me to network with science communicators which ultimately led to my first job in the field.

TIP: If you are unsure of what career you want to pursue or with whom to network, begin by following your interests. There may be an actual job that exists that encompasses those interests, or you may just end up having to create the job yourself.

If, however, you know exactly what career you want right now, you can craft a networking plan that ensures you talk to people in the discipline of your choice to start making inroads into those networks and networking nodes, to position yourself in front of decision-makers, to amplify your brand and reputation, and to find out about hidden and non-hidden career opportunities. Specifically you want to learn the following:

- What jobs exist in this field?
- What career-making opportunities lead to these jobs?
- What are the entry points to these opportunities?
- How do people find out about opportunities?
- What skills are needed for these positions?
- What assignments are required?
- What is the organizational/vocational culture surrounding these careers and jobs?
- What kind of résumé/application is required?
- What are the deadlines for the job and career opportunities?

But even if you think you know what you want to do with your life now and in the future, and you have your heart and mind set on a "dream job," you have to

TIP: Build a career plan that allows for contingencies and flexibility.

build a career strategy that allows for contingencies. I call this the "Astronaut" Syndrome. At some point in almost every nerd's life, you probably wanted to be an astronaut. I know I did. But how many actual astronaut slots are there in any given year? The openings are extremely limited. Does this mean that if your aim is to become an astronaut you should give up on your dream? I don't think you should – in fact, I think you should hunker down and do what you can to achieve it. But at the same time, I think you should be completely realistic and build a parallel career scheme that allows you to pursue an alternate career (or even set of careers) should you find that becoming an astronaut is not possible.

As you ponder your career goals, it is also very useful to consider what you *want* and *need* out of your career. This is also a valuable exercise when evaluating opportunities you should pursue, including those that afford you the chance to expand your networks. For example:

Needs:

- I need money (you have to be honest and realistic about how important this qualifier is to you when selecting a career and a location to practice it in).
- I need flexibility (for family, personal life).
- I need to live in City X (family, etc.).
- I need certain benefits (health insurance).
- I need time/resources to work on my projects.
- I need time to devote to my outside interests (religious, hobbies, and charities).

Wants:

- I want to live in City X (because it would be fun, I like the lifestyle and culture).
- I want to utilize A or B skills.
- I want to research volcanology.
- I want to work in industry vs. academia.

> **TIP:** You are in control of your career. You and you alone make the decisions.

Being able to discern your needs from your wants will aid you in selecting the right opportunities that will help you achieve your career goals. You also need to reflect on what else might influence your career decisions. For example, the state of the economy, your family needs, and hiring trends could all potentially impact your decision-making process.

Finally, you are ready to put your career goals to paper.

- Identify major goal and minor goals (based on needs, wants, loves, and loathes):
  - Major goal: To get a job in field X, Industry Y, in Region Z.
  - Minor goals: To work in the NYC area, with a focus on teaching.
- Decide on deadlines, timelines, milestones, for example:
  - Goal: To get a job, deadline – 6 months.
  - Timeline: Each day, week, month, what will I do to achieve my goal?
  - Milestones: In month 1, I will accomplish X, in week 1 I will accomplish Y, etc.

Analyzing and pinpoint what network(s) and networking nodes could help you achieve this career goal

Some of these networks will become clear to you as you work out your career plan. Some networks will be very obvious. For example, if you were a chemical engineer and wanted to work in the oil and gas industry, you would think about infiltrating networks that give you access to these industries, such as:

- Industry associations, like the American Institute of Chemical Engineers (AIChE), American Chemical Society, Society of Petroleum Engineers.
- Conferences, like those sponsored by the groups above.
- Companies, like BP, Chevron, ExxonMobil, Dow Chemical.
- Regions that play host to these companies and jobs, like Houston, New Jersey, Gulf of Mexico, and the Middle East.
- Groups on Social Media, such as LinkedIn, Facebook, Twitter.
- Alumni from your academic institutions and department.

But even if the networks are not completely apparent, you can still start to reach out to organizations to connect with people who may be in possession of information and inspiration that can steer you in the right direction.

One of the easiest networks for you to access as you start high-impact networking is that of your advisor or mentor. Make an appointment with this professional to discuss your career goals and ideas, and in the meeting, ask who they know who might be able to give you some guidance. Many students and postdocs don't think to pursue the networks with which their advisor is associated, which can and probably does include networks in and beyond STEM. And even if you have already left academia, if you had a good relationship with your departmental advisor, you can set up a phone appointment to reconnect and discuss this. If you don't have a connection with your alma mater, start with your professional mentors.

Keep in mind that although this is one of the first steps in developing a networking strategy, it is also one that is dynamic, in the sense that as you talk with more professionals you will learn about new networks to access. So this stage almost forms a continuum – the more networking you do, the more networks and networking nodes you find out about and the more networking you can do in those networks.

Establishing networking goals to access the network(s)

These objectives are dependent upon your career goals. If you career goal is to get a job in 12 months, then you can establish an initial networking goal of connecting with two new people each week, via cold emails (see below). As you get closer to your goal deadline, you can modify your actions to connect with four or five new people each week.

To get started, consider this as a framework for your networking goals:

- A good daily goal: Identify at least one new person you would like to contact.
- A good weekly goal:
  - Send out 2–3 cold emails to people you have not spoken with before.
  - Send out 1–2 emails to people in your network with whom you haven't spoken to in a while (at least a quarter of a year).
- A good monthly goal:
  - Arrange to have an informational interview or informal conversation with 1–3 new people a month.
  - Look for at least one new network or networking node to access, or one opportunity to network – such as a new conference, a professional society, an article, or a local charity organization or event.
  - Look for one opportunity to self-promote, such as applying for an award, listing an achievement in your alumni magazine, or offering to write a blog.

Finding contacts in the network(s)

Chapter 6 is devoted to helping you identify people for your networks. But I want to share with you the key to unlocking any contact.

Leverage the theory of six degrees of separation. As I mentioned in Chapter 1, six degrees of separation states that we are connected to everyone else on the planet by no more than six degrees. A degree is defined as a shared relationship, and thus I know someone who knows someone who knows someone who knows President Barack Obama. This is an extremely important concept in networking, because it proves that the more people you know, the more people you have access to. So if you want to work in the petrochemical sector and you don't know anyone personally who works in that industry, you probably know someone who knows someone else who is a Shell Oil employee, for example. LinkedIn is a social media site that is based on six degrees of separation. As I discuss at length on Chapter 8, LinkedIn actually shows you how many degrees away you are from another professional. The important point here is this: With every conversation you initiate and with every new connection you make, you are getting closer to your goal of accessing the networks of professionals who are the decision-makers in your desired industry and who can offer you access to the Hidden Platter of Opportunities.

Initiating contact requesting "informal conversations" or "informational interviews"

As you identify people with whom you want to connect, your next step is to actually initiate contact.

You always want to initiate contact via email as opposed to the telephone if at all possible. If you call a stranger, with the positive intention of exploring a partnership, you have no idea what that person is working on at that moment. He could be in the middle of a meeting or an experiment, or he might have just had a fight with his boss or his partner, and the last thing you want to do is appear intrusive. So endeavor to find the person's email address to start the correspondence in that way. However, if you don't have their email and the only way you know to get in contact with them is via the telephone, your initial dialogue should be brief and to the point:

"Hi, Dr. Connor? My name is Alaina Levine and I am a volcanology postdoc at the Jet Propulsion Laboratory. I recently read your paper in *The Astrophysical Journal* about X and was interested in speaking with you about it. Would it be possible to make a phone appointment to have an informal conversation about a potential collaboration?"

Note that in this scenario:

- I did not start talking Dr. Connor's ear off. Instead I asked if I could make an appointment. Another tactical phrase you could utilize is: "Do you have a few minutes?" or "Is this a good time to chat for a short while?" Both statements demonstrate respect for Dr. Connor's time and don't interfere with his day. And if he answers that "yes this is a good time to speak," your next statement can go into some further detail about why you are interested in his work, and what you hope to gain from a conversation.
- I used the phrase "informal conversation" or "informational interview." Both communicate to the other party that you are not looking for a job and you are not trying to *get* something from them. Instead you are honestly hoping to exchange information and explore the potential for partnership.
- I delivered an extremely abbreviated version of my brand statement. I could have added a few extra details about my area of expertise, but in the first few moments when you call someone, you often catch them "off guard," in the sense that their brain doesn't register who you are at first. So your best bet is to keep your brand statement short and sweet.

TIP: The phrases "informal conversation" or "informational interview" are code for "I don't want to extract something from you; rather I want to explore the potential to provide value to you and vice versa."

In the event that you do have the person's email address, your initial "cold email" (correspondence with someone you have never met before) could follow this format (this comes from a real email I sent – see Chapter 6 on the section pertaining to Articles for the story behind this email):

Greeting: "Dear Ms. X,": Note that since this was a cold email and I did not know her personally nor was I referred to her by someone else, I used the formal "Ms." to address her. If she was an academic, I would have addressed her as "Dr."

- An introduction of who I am: "My name is Alaina Levine and I am Director of Special Projects at the University of Arizona College of Science."
- What prompted me to reach out to her: "I read with great enjoyment your article in PR Week entitled 'XX'."
- Why I liked her article: "I found your article extremely intriguing, because A and B."
- Why I wanted to speak with her: "I am a public relations practitioner and part of my current position involves a mix of media relations and outreach to children and adults. I would be very interested in speaking with you about your work and XXX."
- The hook: "I have never worked for a PR firm before and would be very interested to learn your insight into transitioning into this sector with my public relations experience in academia."
- An action item: "Would it be possible to arrange a short phone appointment to discuss this further? As I am based in Tucson, I am on Pacific Standard Time. Please let me know what dates and times in the next few weeks would work best for your schedule."
- An expression of gratitude: "Thank you again for writing this clever article and for the opportunity to speak with you about it. I look forward to hearing from you and to discussing it with you soon."
- Ending Greeting: "Best regards."
- Signature: Alaina G. Levine, Director of Special Projects, College of Science, The University of Arizona, phone number, email, website.

Another format:

To: Dr. Nathaniel Connor
Re: Referred by Dr. Richard Feynman, Caltech

Dear Dr. Connor,

My name is Alaina Levine and I am a postdoc in volcanology in the laboratory of Dr. Richard Feynman at Caltech. My work centers around X and Y. Dr. Feynman suggested I reach out to you as there seems to be some <u>synergy</u> between our interests and I would like to <u>explore a potential collaboration</u> with you.

I found your recent article in *Physical Review Letters* to be fascinating. I appreciated what you had to say about A and B and was curious how you made the conclusion about C. In my current experiment, I am analyzing A and B+1 and found that C only came about when we added D.

Given the similarity of our work and goals, I think there might be an opportunity to partner and I would be very interested in speaking with you about this. Would it be possible to arrange a short phone appointment in the next few weeks to discuss this? Please let me know what dates and times would be most convenient for an informal conversation.

I would be happy to send you my CV for review, but meanwhile, I invite you to visit my LinkedIn profile at www.linkedin.com/in/alainaglevine. Thank you very much for your interest. I look forward to speaking with you and to perhaps working with you in the future.

Best regards,
Alaina G. Levine
Postdoctoral Associate, Volcanology
Department of Physics and Astronomy
Caltech
520.555.1234

Note the key phrases relating to synergy between our interests (a common thread) and an invitation to explore ways to contribute value to each other. I also didn't attach my CV there but rather offered it to him if he so desired it.

Conducting the informational interviews

The informational interview is the lifeblood of the networking strategy. You won't be able to gain any new information or mutually beneficial partnerships with people without an informational interview. The aim of the informational interview is simple and clear:

- To open the lines of communication …
- To exchange information …

- To plainly elucidate and understand each person's brand …
- To learn how you could craft a win-win partnership …
- To solve each other's problems.

An informational interview is simply a conversation you have with another party that is designed to exchange information between the two of you. Most people in the business world know this phrase and the concept behind it. But scientists and engineers might not know it, and you should because it is an important tool in your networking and career exploration toolbox.

Unlike a job interview, which is designed to assess whether a candidate can solve the problems required of the position and add value to the organization, an informational interview is designed to bring to light information between and about two people. It is sincerely meant to be an exchange of information, in that you are providing something just as much as the other person is. The goal is to determine whether the two of you can help each other with your problems, something that cannot be determined until you have the informational interview.

Every time you initiate contact with someone, you should ask for an informational interview, or an informal conversation. The person to whom you are sending this request is going to see these words and most likely they are going to respond positively. Why? Because the phrase "informational interview" is a code word in the business world that clarifies you:

- are not desperate for a job from this person,
- are not trying to extract something from this person, and
- don't need something from this person.

Rather, you want to have a conversation to exchange information – they may provide information about their field, organization, career path, and so on, and you can provide information about how you might be able to help them solve their problems.

TIP: Everyone, no matter their field, sector, job, or station in life, is always looking for someone to help them solve their problems.

Recall the conversation I had with the woman and her awesome high heel shoes on the airplane: We had an informational exchange. We talked about her work, her needs and her goals and I talked about my work and what I could potentially do for her that would solve her problems. It wasn't a job interview. It was an informational interview and it provided value for both me and her.

Now of course when you do an informational interview I realize that often, especially if you are currently unemployed, you may actually feel like you want to get something from the other party – namely a job, or at least an interview or access to decision-makers working within their organization. So you may feel desperate. But you don't have to demonstrate that to them. Because when you use the phrase informational interview in a cold email the other party is more likely to respond to your query than if you send me a frantic-sounding email pleading for a job or an opportunity to interview as a postdoc in the lab or a manager in the department. So to be clear, the informational interview phrase sets the tone of your conversation and your needs – you want to learn more about what they do and explore ways in which you can add value and solve their problems.

What the heck do you talk about?

One of the biggest stumbling blocks that many neophyte networkers encounter is determining what the heck they should speak about when they are meeting someone new either for the first time, like at a conference, or have arranged an informational interview via a cold email.

Do some research in advance. Remember the job search is never about you – it is always about what you can do for me, the decision-maker. As such you have to do research on the other party, their organization, and so on in advance of the informational interview. They need to see that you are invested in the conversation and the relationship with the other person – that you are not just out to get something from them, but are truly interested in contributing.

> **TIP:** The job search is never about you, the job seeker and what you can get from me, the decision-maker. It is about what you, the job seeker, can provide me, the decision-maker.

They want to see that you have thought this out and through thoughtful research have come to realize that your expertise very much aligns with the needs of the organization and you want to explore how you might be able to help them. This is the "synergy" to which I referred in my email above. You can find information on the other party via:

- A basic Google search.
- A LinkedIn search: Examine their profile.
- If they have a personal website, like if they are associated with a university or other academic institution, take a look at what's on their site.
- A "Google News" search on their name – have they been quoted in the media recently? Did they just win a big award or grant? Did their institution issue a press release about their accomplishments, or that of their organization?
- A journal search – have they just written a paper?
- A search of relevant conferences – Are they going to be or have they just given a talk at a major meeting?

These are a few ways you can obtain vital information about the person on which you can base your informational interview.

So you've now done your research about them and can start to assemble some questions you would like to ask. These may include:

- What's the best part of your job?
- How did you find or carve your career path?
- What skills are the most useful to you in your current position and throughout your career?
- What skills do you wish you had had when you began your career or current job?
- What are some entry points for a career or job path that you have?
- What kind of information or experience should I especially highlight in my cover letter/résumé/application?
- What are some of the biggest mistake applicants make when applying for or pursuing job opportunities in your organization/field/department/industry?
- Is there a typical day or does each day vary? Can you describe a day?
- What are some of the proudest accomplishments you have had in your career/current job?
- What advice would you offer someone who might want to transition into a career path similar to yours?
- What organizations/associations are most useful to be members of for your career?
- What conferences are the most useful to attend?
- What publications are the most useful to read?
- What LinkedIn groups are the most useful to belong to?
- Who else should I be speaking with about these topics?
- What is your mailing address? (So you can send them a thank you note.)
- If there is anything I can assist you with, don't hesitate to let me know!

Keep in mind that although you are asking a lot of questions during the informational interview, it should not feel to the other person like you are frenetically peppering them with queries. Rather it should be a conversation. So you can feel free to inject information about yourself and your own expertise and interests into the conversation.

And of course these questions are simply the skeleton of a great chat. But don't feel like you have to stick to a script – if someone says x and that gives you an idea about y, then go on a tangent and talk about y. Furthermore, the informational interview works both ways – just as you might get ideas from speaking with them, they may get ideas from speaking with you.

I want to highlight a few of these questions.

- What's the best part of your job?

> **TIP:** As you participate in informal conversations, you can inject information about yourself that is relevant to the discussion, in part so it doesn't feel like you are peppering them with question after question.

As I discussed in Chapter 1, this is one of my favorite questions to ask and I often use it very early in a conversation when I first meet someone. This sets a great tone for your discussion and ultimately for the alliance you are hoping to build as well. And this question is almost foolproof – it gets people talking about themselves – as noted above – and gets people talking about what they enjoy and as they equate that joy with you, they are more apt to reveal more vital information and knowledge to you.

Of course related to this subject, you never should say something like: "Oh you work for University Y? Oh that is such a horrible place. Why on earth would you work for that organization?" You never want to invoke anything but pleasure especially in the early part of a relationship. It's like dating: You keep the other party engaged with a conversation about what makes them happy.

- What skills are the most useful to you in your current position and throughout your career?

  This is a great question because it allows you to learn new information that can set you on a path for success within this sector, organization, and field. It gives you insight into what abilities you should concentrate on, sharpen, or highlight on your résumé or in your cover letter. Additionally, you can leverage the question as a channel to share your own experiences. So if they say that budgeting has really helped them as a faculty member, you can respond by agreeing with them about its relevance and mentioning you are developing budgeting and accounting skills through your volunteer work in the postdoc association at your university or as a member of the board of directors of a local charity.

- Who else should I be speaking with about these topics?

  You are always seeking to expand your networks, so this is a critical question to ask. And the beauty of this question is that almost always the person will give you the contact information of at least one or more people for you to pursue.

  And as you develop contacts with these other professionals, always remember to circle back around to let the original party know that the referral was very helpful for X reasons (such as a great conversation or you got a job). They should know that they helped you and you appreciate that assistance. It also serves as another avenue to keep the flow of information continuous.

- What is your mailing address?

    The purpose of this question is simple – you want to send them a thank you note and you need to confirm you have the right mailing information. See below for details on the art and science of the thank you note.

- If there is anything I can assist you with, don't hesitate to let me know!

    This statement/inquiry also serves to set the tone of your future relationship. You want to ensure that they know you are not out to get something for yourself but that you see this as a mutually beneficial partnership. Now they have just donated time to you – they spent 20 or 30 minutes or perhaps more out of their day talking to you, and this is time that they could have used to devote to their scientific and engineering exploits or to playing ball with their kid. So in addition to thanking them on the phone and then sending a follow up thank you note/email, you also want to express gratitude in a different way by telling them if they need anything they should simply contact you. And this also clarifies your resourcefulness. You can insist: "Even if I am not the right person or don't have the answers to your queries, I will find someone who is." People really appreciate this and it shows that you'll go the extra mile to help them with their projects. This is a coveted trait that employers look for in making hiring decisions and now you have just demonstrated it to someone who has the ability to offer you access to hidden career opportunities, or can pass your information on to another colleagues with a glowing recommendation (advancing your reputation).

The Art and Science of the Thank You Note

Although it may be a foreign concept for scientists and engineers, the idea of sending a thank you note to express gratitude for someone taking the time to speak with you is extremely important. It can literally build and solidify relationships and can be the action that immediately sets you apart from everyone else who is jockeying for attention from a particular person. Twelve years ago, I attended a conference in New York. After the speaker completed his address, I approached him and introduced myself. I thanked him for his talk and asked him a short follow-up question. Then I asked him for his business card and inquired if I could contact him again about this subject at a later date. He said yes. On the plane ride home, I wrote out a thank you note on stationary. I sent it to him within 24 hours of landing, and then something remarkable happened: Within a few days, the gentleman emailed me to thank me for sending him the thank you note. That one conversation and the thank you note set the tenor for our partnership and as a result he gave me ideas and access to new networks that I never would have even known about, let alone actually been able to infiltrate.

That is one example of how a thank you note helped seal the deal of a networking partnership, but to be frank it is not the only time this has happened to me. I send thank you notes, on stationary and in the mail, all the time, and I often receive emails thanking me for my thank you notes. Why is this? The reason varies:

- in this day and age, very few people know to send notes of gratitude at all, whether on paper or via email, and even fewer do so on stationary;
- very few people realize that the act of expressing gratitude is an important element of being a professional, no matter your vocation;
- very few scientists and engineers expect to receive thank you notes, so when they do it is a welcome and appreciated surprise.

> **TIP:** Know the culture of your field, industry, and geographic region so you can send the most appropriate thank you note.

Remember, when someone takes time out of their day to spend even 5 to 10 minutes with you, discussing career options, or the potential for collaboration in some other form, they are literally losing two precious resources which they may never be able to reclaim: Time and money. The loss of time is obvious, but in science and engineering many may not realize the loss of money as well. Even as STEM professionals, we lose money when we are not completely focused on our technical task at hand – time is money, after all.

Thank you notes are another element of typical business school curricula; business students, especially those in sales (who want to be continuously successful in sales) know they need to send thank

> **TIP:** There is never a negative connotation associated with expressing gratitude to someone for helping you.

you notes to clients and potential clients. But scientists and engineers are not taught this action and, furthermore, since many scientists and engineers don't view their endeavors as being related to a business enterprise, they wouldn't consider a thank you note as part of the cultural norm of their discipline or industry. Being able to know the culture of the organization and community you wish to access or contribute to is important because it helps you determine what kind of note to send. And take "note": I didn't say whether to send a note or not. You should always send a note. But different cultures dictate different ways of doing so.

For example, I recall speaking to postdocs and graduate students at a specific STEM conference a few years ago about networking and I pointed out that they should not neglect to send a thank you note after having a "meaningful engagement"; in this case, I meant anything from having coffee with a new colleague at a conference to an informational interview to a job interview. A number of the scientists immediately protested and stated that they would never "bother wasting someone's time" with a note of gratitude, and that such an action would demonstrate only a negative gesture, specifically that they were trying to "suck up" to the other party.

I explained this was faulty thinking. The other party, whether they are the head of a prestigious laboratory or observatory, or are an early-career scientist themselves, is (for the most part) a human being. And as such, a human being likes to be thanked for helping another human being out. Yes, it is that simple. So by expressing appreciation you are doing the most basic action a human can do when aided by another.

> **TIP:** Since very few science and engineering professionals send thank you notes, by doing so you immediately amplify your brand, attitude, and reputation and put yourself in an advantageous position with the other party.

Depending upon the circumstance and field of science and engineering, an email thank you note may suffice just fine. In fact, in some circumstances I would only send an email thank you note. For example, if the other party works for the United States federal government, I would use email to express my gratitude, since post is unreliable due to increased security measures. In addition, if the person was a high-level leader in their organization, such as the CE of a company, they might not even open their own mail. In this circumstance, I would send both an email and hand-written note, just to be on the safe side.

But what about a job interview? What kind of note should you send in this situation? Again, this depends. I would advise sending an email note within 24 hours of the interview. This is an almost foolproof way of expressing gratitude and you want to ensure that the interviewer knows how committed you are not only to the job and organization, but to the elements of professionalism that are associated with your industry. So send an email note quickly.

But sometimes, I also send a hand-written note as well, even for job interviews.

If it is an informational interview, whether on the phone or Skype, or in person, I often send both. In the email I go into more detail about our discussion and I follow up on points we may have discussed in our meeting. In the hand-written note, I simply express my appreciation and share that I am looking forward to working with the party in the future.

Some tips about thank you notes:

- Always get and confirm your contact's correct mailing address and email address.
- Take stationary and stamps with you when you travel (especially to conferences and job interviews). I often end up writing my thank you

cards on the airplane home and then mailing them in the airport's mailbox.

- Don't forget to include your business card in the note. Even if the other party already has one, it doesn't hurt to include one more that they then have handy on their desk.
- Keep track of your notes. Just like all of your networking activities, it is always a good idea to keep track of to whom you sent thank you notes, what form they took (email or snail mail) and when you sent them. You can keep track of this in your contact management software and make a note of it as it is associated with the contact, but I also find it helpful to simply make a note in my dayplanner concerning to whom I sent thank you notes that week. This allows me to keep track of what kind of correspondence I have with collaborators, which is especially important at the beginning of the relationship, when every interaction you have with the party is critical to the success of the long-term relationship because it sets the tone of the relationship. You also don't want to forget that you already sent a note and accidently send another one!
- Send a note immediately, like within a day or no later than a week if possible after a meaningful engagement, such as an informational interview.
- Send a thank you note after someone does something significant: You shouldn't only send notes after informational interviews. If someone writes a letter of recommendation on your behalf, or refers you to a job lead, or nominates you for a leadership position on a committee, express gratitude in written form.

What to write:

> Dear Sydney,
>
> It was a pleasure speaking with you today. I appreciate you taking the time to chat with me about your experiences in nuclear waste management. After our conversation, I am even more excited about the possibility of contributing to your organization/collaborating with you. Thank you again for the opportunity to speak with you and I look forward to working with you in the future. If I can assist you with anything, don't hesitate to let me know.
>
> Best regards,
> Alaina

Following Up

So you meet someone at a conference, you have a coffee or Skype appointment with them, you send a thank you note, and then what? This next part – the part in which you expand beyond the scaffolding of a relation-ship and actually build it already – is a challenge for most people. After all, once you have that first encounter, what else can you possibly say?

First of all, after you have that first meaningful engagement, connect with them on LinkedIn. (More information on this in Chapter 8.)

Next, you will want to make a point of building a schedule that ensures you continuously stay connected with your networks. One of the tactics I employ after I converse with someone is I make a note in my calendar about 3–4 months ahead, reminding me to contact them again.

When you do follow up, your correspondence should encompass the following items:

- Deliver something of value: Information that can assist them with their problem-solving, or something that is relevant to their profession, industry, or organization, or something that relates to their success.
- Share new data about yourself: Information that can inspire them to give you access to hidden opportunities or new networks.
- Give them an action item: Request that they respond to your note.

Conduct research on them and find out what is new – you always want to deliver something of value to them. Potential ways of doing this include:

- Offer congratulations: If they just won an award or a grant, or just got a new job, or just published a major paper, send them a congratulatory email. Very few people do this, but the fact that you took the time to do so demonstrates your professionalism and your genuine interest in their career.
- Offer new sources of inspiration: I call this the "I thought of you" email, because if I find out something that may be pertinent to my contact, I will let them know about it. For example, "I read this article (or saw this talk at a conference) and thought of you – this reminded me of your work and I thought this was of interest to you."
- Inquire about their work: "I read your new paper in *Cell* and found it fascinating for these reasons – what was your main research method-ology? How did you come to this conclusion? Did you consider using data sets from X institution/experiment?" Asking these questions engages them in a discussion about their work and also demonstrates your interest in their professional success. Just don't ask questions that require too long a response, because the other party may view it as a hassle, as opposed to a welcome conversation.
- Share news about yourself: "I have good news and wanted to let you know about it" – even though this is not exactly about them, it shows

them that you want to stay in touch and it gives them more information about you and the value that you can provide which can unlock hidden opportunities for you to collaborate. Furthermore, your good news may impact them in a positive way. For example, if you just landed a new job at Company X, you can share this information and suggest that there may be an opportunity to work together in this new capacity.

- Ask for their advice: If you just got a new grant or got elected to a leadership role in a professional society, for example, you can share this good news and then say something like "do you have any advice for me as I take over this position/grant?"

In fact, with every follow up email, don't hesitate to mention something about yourself. Perhaps you just published an article or got a new job or spoke with that person's colleague at a conference. All of this information about you stimulates them to ponder novel avenues for collaboration and solidify and grow the relationship. Remember – the more information they have about you and your interests, goals, and expertise, the more opportunity you create for hidden opportunities to be revealed to you or for the other party to spontaneously realize that an opportunity exists or to create one just for you.

You can also send notes to your associates, via email or on stationary, when they do something extraordinary, like win an award, or have achieved a milestone like a new job.

And there is one other form of communication I recommend employing for following up with contacts: Sending a holiday card in December. Keep in mind that the card should not specify a religion (so don't buy a bunch of cards that state "Merry Christmas"). Instead a generic "Holiday Greetings" or "Seasons Greetings" card can really go a long way to help you stay connected with your networks. I send holiday cards every year and yes, I hand write them for those people for whom I have mailing addresses. If you only have their email address, then you can send an email holiday greeting to your contact. Some tips associated with this action:

- If you send an email, don't send a batch email. Yes, it takes extra time to copy and paste a message into 400 contacts, but believe me it is worth it. Of course various software packages will do this for you.
- You don't have to spend a lot of money on holiday cards – in fact, you can buy them at significant discounts usually the day after Christmas.
- The message that you write can be very simple, such as:
  - "I hope you have a lovely holiday season"
  - "Happy Holidays and Happy New Year"
  - "Enjoy the season and I look forward to working with you next year"

- Some businesses when they send holiday cards add a sheet of paper with some information about the person or the business over the year. I personally haven't done that before, although that may change in years to come.

Organizing Your Network

Way back in the last century, when I was a novice at networking and didn't go to a lot of mixers, I kept track of my networking using a single notebook. Each page in the book would be dedicated to an event and I would designate separate columns for the name of the person I met, their organization, a few words about what we talked about, and dates of any follow up actions I took.

This was all fine and dandy if I was only networking at events and only went to a few a year. But of course, that's not how your networking strategy should play out and mine certainly didn't. Pretty quickly the notebook method became useless and obsolete.

So I started using a software program called ACT, which allowed me to keep track of contact data, letters and emails I sent to people, and even notes from conversations I had. Although I don't use this system any more, I found it to be effective when I was just beginning to do high-impact networking and needed to keep track of as much information about parties as possible.

Today there are plenty of options on the market. There is software that you can buy that is integrated with hardware that can scan business cards into your own contact directory. You can use systems that are stored on the cloud. You can probably just get away with using Outlook and other free systems available on the internet. The bottom line is that not every contact management system will work for everyone, so you have to find a system that works best for you. Today, my personal system consists of a combination of my email platform, LinkedIn, and my dayplanner.

Continuing to Follow Up and Look for Opportunities to Exchange Value Over Time

As I have mentioned several times, networking goes beyond correspondence. And just as you are hoping to gain access to the Hidden Platter of Opportunities from your contacts, they too want something of value from you. You have access to networks, inspiration for problem-solving, and even your own Hidden Platter of Opportunities that you can always share. Endeavor to always look to contribute, to always help the other party in some way. They will appreciate it and it will deliver you both significant returns on your networking investment.

Chapter Takeaways

- Develop a networking strategy and plan that is realistic to accomplish and is flexible enough to adjust as needed.
- Identify and incorporate your career goals into your networking strategy, and align the two.
- Seek to find and access networks and networking nodes that can aid you in your career goals.
- Find contacts for your networking by conducting research on trends and news in the fields you hope to access.
- Initiate informal conversations with contacts to begin the exchange of value.
- Express appreciation for those who take the time to speak with you by sending thank you notes.
- Follow up, stay in touch, and organize your networks to continue delivering value to your contacts and exploring opportunities to partner.

6

Identifying People for Your Networks

Identifying people for your networks can be tricky, especially as you contemplate career moves outside your current discipline, sector or even geography. If you are vaguely interested in working for a museum in San Diego, how do you start "networking" to access people in that field and region?

Start With Who You Know

No matter what your networking goal is, you should always start by asking people with whom you already have a connection about other potential collaborators. As I mentioned above, this can include your advisor or mentor, but don't limit yourself to only people you perceive are in your field. For example, your friends and family may not be in STEM, but they have their own networks full of people, including those with whom you could directly exchange value and those who are connected to others (through six degrees of separation) who could offer you access to hidden opportunities. For example, I was brainstorming with an engineer about possible networking leads when he suddenly remembered that his former rabbi lives in a town that is home to a major manufacturing facility in the engineer's field. He recognized that this spiritual leader would have access to other engineers who are employed by the plant as well as ideas for hidden opportunities for collaboration, given his connection with the community. At the same time, I realized that my high school buddy works for a company that this engineer had an interest in, thus I put the two of them in contact with each other.

Don't be shy about asking your family members and pals about people they may know in their own circles, through their employment, community, philanthropic endeavors, and religious affiliations. And of course you can always chase down friends with whom you haven't spoken in a while, such as chums with whom you went to college or graduate school.

Networking for Nerds: Find, Access and Land Hidden Game-Changing Career Opportunities Everywhere, First Edition. Alaina G. Levine.

Since you are looking to network – to craft a win-win partnership – and not extract something from them, you don't have to feel timid about making contact or that you are putting them in an uncomfortable position. Just make it clear from the beginning that your aim is to reconnect, learn what they are working on, and explore the potential to perhaps provide value to each other. And quite frankly, because everyone is looking to solve their own problems in new ways and networking provides a means to accomplish this, your outreach to them, when appropriately framed, will be appreciated.

Institutional Networking

You would be surprised at the plethora of networking opportunities that exist in your current institution, no matter your industry. In academia alone, you could pursue networking through any of the following channels:

- Your department, such as student clubs, journal clubs, departmental advisory and alumni boards.
- Your college, such as industry advisory boards, alumni boards.
- University, such as the postdoc association, or the graduate and professional student council, student alumni board.
- Clubs associated with hobbies, culture, nationality, religion, and politics.
- Clubs associated with philanthropic endeavors.
- Clubs, boards, or groups associated with diversity, such as a women's advisory organization or the LGBQT society.
- Organizations devoted to professional development of various institutional constituents (faculty, staff, postdocs, students).

You could find out about these different organizations, contact the chair or head and ask about becoming involved or attending their events. If they are formal associations, you could potentially learn about joining and running for office (see more detail below in Professional Societies). If they are boards that don't offer membership, such as an Industry Advisory Board, you could approach the university liaison (who could be the unit's development director or the head) and ask if you could present your work to the Board and attend their receptions.

If you are at a major research university that has a business and/or law school, you will also likely find some terrific networking opportunities existing within these units. More and more business and law faculty and program directors are recognizing the value of engaging scientists and engineers, especially when it relates to entrepreneurial enterprises. When I was at the UA, the business school regularly hosted events that engaged

> **TIP:** One of the hidden jewels of networking within universities is the development director.

the local industrial community, and enlisted STEM faculty, students, and postdocs to contribute to business research and projects. These activities delivered extremely high networking ROI, in that they attracted a diversity of thought leaders in myriad fields who were extremely enthusiastic about potentially joining forces with scientists and engineers.

One of the surprising and often overlooked networking nodes that exist within an academic institution is the development director. Every college and university has someone whose mission it is to raise funds; in fact, the larger the institution the more development directors they have. If you approached them and volunteered to assist them with their endeavors, by presenting your research to high-net worth individuals, accompanying the director on meet-and-greets with local companies, participating in outreach events, or even giving them access to your own networks, they probably would greatly appreciate it. After all, they are trying to raise money to support you and your research, so who better to communicate the importance of your investigations than you? And investors love meeting the people they support, because it gives them a chance to observe the positive impact of their money. By aligning yourself with the development director, not only would you be able to network with all of the people in their network (which is extensive), but you would also gain an opportunity to sharpen your communications and marketing skills and promote yourself in an appropriate manner. This partnership is a prime example of a win-win for everyone, and is extremely underexplored by scientists and engineers.

Additionally, if there is a particular career path in which you are interested, why not see if there is anyone who works for your institution in that position or profession? Consider the following career interests and the correlated professionals or departments that exist in your organization:

- Journalism, communications outreach: Speak with the public relations, media relations, community outreach units.
- Technology transfer, patent law, and entrepreneurship: Speak with the office of technology transfer, entrepreneurship program.
- Policy: Speak with the office of governmental relations.
- Scientific problem-solving: Different from research, scientific problem-solvers can work in various roles in academia, from managing labs to designing and building specialized equipment, such as telescopes and sensors. At the UA, for example, there is an entire cross-disciplinary team of optical scientists and engineers who support research projects in astronomy, planetary science, and optics by lending their expertise to solve technical problems that advance the research goals of professors.

Finally, follow your institutional calendar very carefully, because there are always events that they sponsor or in which they are involved that could provide you with networking channels.

Note: Even if you are not at an institution of higher education, fear not: Many of these options are mirrored in companies, government laboratories and agencies and even non-profits.

Diversity Groups Within Organizations and Companies

Large institutions and companies, particularly those that are multinational, often have diversity organizations within their borders. This is not because the companies want to be philanthropic; on the contrary, these businesses recognize the critical importance that a diversified workforce provides to the bottom line. So they often support organizations that focus on specific diversity characteristics. For example, Raytheon has a number of diversity-focused organizations, such as the Raytheon Women's Network and the Raytheon Hispanic Organization for Leadership and Advancement. If you are looking to network with someone from Raytheon, one tactic that you could employ would be to contact someone from these organizations.

TIP: When researching a company, look into their diversity organizations and contact the chairs. You can often find this information via a simple Google search.

The diversity groups exist to support the advancement of professionals in those communities. They often have their own leadership and networking opportunities, in the form of events, workshops, and conferences. These are just some of the diversity communities that you could tap into:

- women
- Hispanic
- African American
- Asian American
- LGBQT
- Native American
- those with disabilities.

If you were interested in a particular company and wanted to connect with someone inside, you could utilize the diversity groups as an entry point. And of course once you are employed by a company, do take advantage of these groups for your own networking, skill enhancement, and leadership development purposes. Often, the elected leaders of these groups have the ear of the company's executive team, giving you even more access to hidden opportunities for networking, advancement, and self-promotion which ultimately can benefit both your career *and* the firm.

Professional Societies

> TIP: Professional societies exist to advance the profession and the professional.[i] They want you to engage them!

Most professionals know about their own professional association, but many people do not consider the wealth of networking and self-promotion opportunities that exist within that organization. Yes, they know about the conference and maybe they read the newsletter, but there is so much more that you can experience from being a member or simply demonstrating your interest in membership.

First and foremost, know this: The society exists and thrives because of YOU. Just as a company depends on customers to buy its products to ensure its stability and continuous growth, so too does a professional organization rely on YOU as a member to ensure its future success. After all, the association cannot function without members – you provide not only financial support (through your dues and conference and publishing fees) but also the proliferation of scholarship itself. As such, the managers and the boards of professional societies have a duty to continuously recruit, retain, and sustain members. They want YOU as a member! And more importantly, they want you to be an engaged member, because an engaged member is one who ultimately expands the breadth and depth of their support in myriad ways (giving more money, volunteering their time, promoting the organization to their own networks, etc.). This gives you an advantage in your networking because as a member, you are a stakeholder, you have a voice, and you have a vote as to how the organization moves in the future. And you should most definitely take advantage of the abundance of networking opportunities being a member affords you.

Tactics:

- Research what societies are relevant to your career and networking goals. You can start by Googling the major phrases associated with your interests, such as "cell biology" and "society" or "museum evaluation" and "organization." You can also use LinkedIn to find the societies of interest. As stated above in the section on Informational Interviews, you can ask people who are contacts in your network such as your advisor (past and present), your colleagues, and even your boss. If there are professionals who you admire, you can take a look at their LinkedIn profile and see what groups they belong to. Similarly in academia, many scholars post their CVs on their websites, and you can review their CVs to see with what organizations they have aligned themselves. But don't limit yourself to the main society associated with your discipline. There are special interest organizations, such as the Society of Women Engineers, Society of Hispanic Engineers, the Society of Black Physicists, and SACNAS (which used to be called the Society for the Advancement of Chicanos and Native Americans in Science), all of which exist to further the professional goals of its members and, as such, provide extremely high networking ROI.

- Investigate membership options and benefits. Different organizations have different membership tiers, depending upon where you are in your career. Often, early-career professionals, or those who are experiencing a mid-career change, can take advantage of lower-priced membership levels. And with your paid membership comes myriad benefits, many of which can lead to their own networking and inspiration opportunities as well. Your objective is to take advantage of as many benefits as you possibly can. For example, most STEM professional associations produce their own journals, and offer skill-building and career-advancement webinars, which are only available to paid members. Furthermore, your society may have its own internal social network platform that can also aid you in connecting with other members and leaders in the field.

  Ascertain the society's organizational structure. Analyze the website and/newsletter to determine how the society is managed and organized. What departments do they have? What committees encompass the society's makeup? Most major societies in the US and abroad have a paid staff whose goal it is to manage the organization, as well as an elected board. The board usually consists of scholars in that discipline, whereas the staff often have professional backgrounds in non-profit management. This is vital information to know because you want to assess who you should be contacting in an organization. You also want to start thinking about volunteering and serving on committees (see below) and you can get insight into how everything runs efficiently via the website and newsletter. It is also very good to know who is running for office and/or who just got elected and who is overseeing or chairing committees. On the other hand, other societies, especially those that are either smaller in size or extremely specialized, may only have one paid staff member – the executive director. They rely on an eager army of volunteers to ensure the association is surviving and thriving, and would certainly welcome you into their clan as an unpaid helper.

- Review their website and other promotional materials. There is a wealth of information you can gain from reading just the newsletter of an organization. I currently belong to three professional societies and I read their newsletters cover to cover every time I receive them (both digitally and on actual paper). Among the valuable insights I gain:

  o The organization's current priorities, programs, and projects and who is running them.

  o What awards the organizations are giving out and who has won them.

  o Accolades for members' professional triumphs, such as honors they have received, books and papers they have authored, new positions they have attained, and patents they have been awarded.

- ○ Financial reports (which have to be made public to all members), which also give insight into society activities now and in the future. All of these outputs arm me with data that help me envision potential partnerships.
- Join an organization, if you haven't already done so. You must do this. A paying member, even if you are a student, has the required currency to interact with the leaders in the group and other members. Believe me, even if the actual price of membership seems like a large amount of money at this point in your career, it will pay off. The networking ROI in joining an organization is invaluable.

   However, I recognize that there might be more than a few professional societies that are aligned with your career and disciplinary interests. You don't have to join all of them – join those from which you envision you will get the most networking ROI. And for some associations, there are "benefits" of which you can still partake even if you are not a member. For example, some societies allow non-members to join certain listservs and access to the newsletter (i.e. it's not behind a paywall nor does it require you to login first to read it). And you can always attend the conference (albeit at a higher non-member registration rate).

- Contact the organization. Start by emailing the "Membership Coordinator" or "Director of Membership" and ask for a phone appointment. Inform them where you are in your career path (I am a grad student, postdoc, new faculty member, or am transitioning into a career in X) and that you are interested in learning more about becoming an "engaged member."They will be more than happy to chat with you – after all, their job depends on building the membership. As you speak with this person, ask them about opportunities to volunteer in the organization, both virtually or in person. Reference and ask questions about items you have read about on the website and in the newsletter or annual report. I guarantee that the Membership Coordinator will be thrilled to speak with you, as very few people in science and engineering take this step to demonstrate their commitment to their society. The simple act of reaching out to them propels your brand and reputation into the limelight. And as you chat, he/she will start to get ideas as to where you might be a good person to serve the organization, what committees are in need, and what help you can provide. In other words, the Hidden Platter of Opportunities will be proffered.
- Access the membership directory. As a member, one of the most valuable benefits you have is access to the contact information of every other member. So if there is someone you wish to contact, you can get their information from the organization. But more importantly, you have a reason to contact them – you want to become a more engaged member and learn from them about how to achieve this. On the flip side, they listed their names and contact data in the directory for a specific purpose – they want to be contacted because they too want to

expand their networks and seek out new career opportunities. You can often search the directory for members by region, company or institution, and even subfields. When you send your initial email to introduce yourself, make sure to write in the Subject line "Member of [Society's acronym] – Appointment to chat", so they know immediately where you got their contact information.

And while we are on the subject, make sure your contact data is listed in the directory and that you keep it up-to-date. Furthermore, if the directory offers opportunities to expand your listing – for example, to add your picture, a URL for your LinkedIn profile or blog, or a section in which you can post portfolio pieces – take advantage of this. The more positive, brand-promoting information that is out there about you, in the real world and via Digital Trails in cyberspace, the more inspiration and enticement you provide for a colleague to contact you and initiate an alliance.

> **TIP:** Don't hesitate to reach out to people listed in the membership directory of your (targeted) association. They made their contact information available because they too want to network!

- Subscribe to communications channels. Get on the mailing list. Join the listservs. Sign up for the organization's social media outputs, such as the LinkedIn group. "Like" them on Facebook, and Follow their Twitter handle and related feeds (see Chapter 8 for more details).
- Volunteer. Volunteer. Volunteer! Wow – I cannot tell you how simply volunteering for non-profit organizations within my field has enriched my career, given me access to multiple Hidden Platters of Opportunities, and helped me craft strategically important networks. Your society leaders will greatly appreciate the fact that you volunteered for *anything*. Volunteers are the lifeblood of an organization and yet very few members take advantage of the opportunity. So by doing so, you promote yourself as someone who is eager, hard-working, dedicated, and invested in the profession. So where and how can you volunteer?
  - Volunteer for committees.
  - Volunteer to write for the blog or newsletter (it doesn't have to be a regular thing – you can suggest one or two articles).
  - Volunteer to work at the conference (see below).
  - Volunteer for their speakers' bureau – not every organization has one of these, but if yours does, consider participating in it. I belong to the Visiting Lectureship Program for SPIE, the International Society of Optics and Photonics, and as such my name and profile are on a special speakers' bureau microsite which is perused by SPIE student chapters all over the world. My brand has gotten significant exposure, and I have been able to profit by connecting with diverse, multi-cultural publics who share the same passion for science and engineering career advancement as I do. And as a

speaker, I then partake of even more promotional, learning, and partnership-building benefits when I visit different campuses. If your professional society does not have a Speakers Bureau, consider volunteering to organize and oversee one, and/or at least helping to get it off the ground. This is a prime example of thinking entrepreneurially about your career advancement strategy, and the experience itself will sharpen your business skills. Your success will elevate your brand and magnify your reputation amongst the leadership. Heck, you might even win an award for your thoughtful service and dedication to the society!

- Apply for awards. Most organizations give out awards on an annual basis. They can be titled anything from a "Fellow" to "Award for Service" to honors in the name of a specific person for achievement in that field. Even established STEM professionals often don't think to nominate themselves or ask others to nominate them for society awards, but this is a sure way to raise your visibility within the organization and discipline and demonstrate that you are a leader in the field. So apply for awards and remember, even if you don't get them the process is worth your time and energy. And furthermore, you can ask to volunteer for the awards committee – they will be ecstatic to have you (that's definitely a tough and time-consuming board on which to serve, no doubt about it) and the networking opportunities there will be extremely useful. And as you build your network, you can always ask others to nominate you for the award too! And thinking entrepreneurially – as you become more engaged in the society, you can always suggest an awards program and help bring it to fruition.

- Seek leadership positions. As you become more entrenched in the organization, consider pursuing leadership opportunities. These can be as informal as a committee chair (which might not even require a vote) to something as formal as a member of the Board of Directors or an elected position such as president or treasurer. As I mentioned in Chapter 4, serving in a leadership role in your professional association elevates your brand, attitude, and reputation. It gives you a reason to reach out and connect with other members and additional society stakeholders. It serves as a credential – the organization elected you to this position, so they perceive you as a leader. Pursuing leadership roles is a sure-fire avenue to high-impact networking and accessing hidden opportunities.

Take note, however, that timing (in various respects) is an important element to consider when seeking a leadership position. You don't want to jump into a committee chair role before you understand the inner workings – the structure, priorities, and even culture – of the association. You also have to consider at what point you are in your career and how much time you can devote to outside committees while at the same time doing your job (or if unemployed, looking for a job). Just like everything else, you have to balance your time commitments to ensure you are productive on all fronts (in part to maintain your stellar brand and reputation). This doesn't mean that you can't or shouldn't pursue a leadership role early in your career – just do so judiciously.

- Take full advantage of their career services. Most major organizations have at least one dedicated employee whose focus is helping members find jobs and career opportunities. Larger organizations often have entire departments devoted to these efforts. Find out what career services your organization offers and take full advantage of them. If you can post your résumé on their job board, do so. If they have a special listserv reserved for career-related discussions, join it and contribute to it. Participate in their professional development and career advancement activities, including webinars, workshops, and teleseminars.

- Join and participate in chapters of organizations. Not every organization has regional or discipline-based chapters, but many do. Chapters often function like mini-versions of the society itself, with its own elected leadership, networking events, committees, and even awards programs. Additionally, they might sponsor their own professional development conference. The real beauty of the chapters is that they often provide networking novices fabulous opportunities to sharpen their skills and promote themselves in their community.

The Public Relations Society of America operates just like this. There is the national association, and then there are regional chapters. When I was employed in a PR capacity with the UA, I joined the Southern Arizona chapter of PRSA and attended their meetings every month, where I networked with other professionals. I gained new knowledge about careers and the practice of PR, and I learned of clandestine opportunities for advancement. After my first year, I volunteered to serve on a committee, which led to me being selected to be the chair of that committee. And the following year I ran for treasurer, which ultimately led to me being elected into the presidential line (president-elect and then president). This significantly boosted my profile and cemented my reputation as a thought leader in the community. It also gave me a pathway to regularly communicate with the national leaders, including facilitating a visit from the national president to speak at one of our chapter events.

The national leaders of an association consider the chapters to be extremely vital and strategic in the growth and progression of the profession (or discipline) and the professionals themselves. Chapters often receive financial support from national, there is usually a local position whose mandate is to liaise with the mother ship, and outstanding chapters are often rewarded with various honors. Research whether your association has a chapter system and what the closest chapters are to your community. And if there isn't a chapter nearby, you can possibly participate virtually with another chapter, or consider investing in attending a regional chapter event in another city. Or perhaps you can assist in launching a new chapter – the national leadership will thank you for it!

- Join and volunteer for special interest groups within the larger societies. Many big organizations have smaller sections within them. These groups can focus on subdiscplines of the main area or field (for example, the aquatic ecology section of the Ecological Society of America), or attributes of its members (such as the Young Professionals Group of the American Institute of Chemical Engineers). Because they are highly focused and smaller in size, they always offer terrific networking opportunities. At the conferences, they often have their own events which you can attend, and they are also always looking for leaders to help plan activities. And if your society doesn't have a particular section that you think would serve its constituency, you can always speak to the organization leadership about launching one.

Conferences

Attending professional conferences is one of the most crucial career-boosting steps you will take along your path.[ii] And yet, most people who frequent conferences do not take complete advantage of the many, many, many (did I mention many?) opportunities they present to grow and solidify your network, learn about your field, discover new arenas, enhance your skills, identify and forge career linkages, promote yourself, and open yourself to myriad other more fruitful and exciting circumstances for professional advancement. The conference is a multiverse of career-augmenting treasure which both emerging and established scientists and engineers should ravenously pursue.

> TIP: To get the most out of conference experiences, strategically plan ahead and be proactive and interactive with speakers and other attendees, before, during, and even after the meeting.

To be clear, attending one session and playing the wallflower at a meet-and-greet is not the way to score conference participation gold. It takes advanced strategic planning and preparation, combined with nimbleness and nerve, to be able to achieve the best return on your conference investment of time, energy, money, and intellectual input and output.

Your planning and preparation will depend on many factors, including your goals for attending and the size of the meeting itself. If it is a major conference where tens of thousands of professionals converge, you will have to develop a very targeted plan to optimize your time and the time of the people you want to engage. But even if it is a small highly-focused workshop meeting, where less than 100 people are attending, the same planning principles apply.

Here are some tactics to help you orchestrate the most rewarding experience at your next conference.

- Start with a goal: Every important move in your career course begins with a goal, and so too does your conference participation. Before you submit the paper and book the hotel, pinpoint from the start why you are attending, and what are the targeted takeaways that you want to acquire or actualize for the experience to be worthwhile. Number 1 on your list should always be to network, as you should constantly aim to increase the number of people you know in your field and in related disciplines. But there will surely be others, and by solidifying your goals for attending, you will be able to form the basis for a course of action to achieve those goals.

> **TIP:** Take advantage of smaller gatherings that occur at conferences, such as business meetings, town halls, and splinter meetings to engage in high-impact networking in a more accessible environment.

- Develop a plan: Your plan will incorporate your mission for going to the conference, paired with specific actions you need to take to accomplish your objectives. It is important to remember that you will have to be completely *proactive* and *interactive* at the conference. In other words, you will have to take control, and reach out to others who are attending. You cannot be passive and hope that networking will just happen to you; you must examine the conference agenda and determine what sessions to attend, AND make a plan to network, introduce yourself to others, and seek to achieve any other goals you have in attending.

- Prepare: After you establish your conference intent, the next step is to prepare. This involves researching who will be attending, papers that will be premiered, sponsoring organizations that will be at the trade show, and companies and universities that will be interviewing at the career center. Make note of smaller, more casual gatherings of like-minded colleagues to hone your networking skills. But also plan to attend open business meetings and award ceremonies. Society "Town Halls" offer a terrific opportunity for early-career professionals to meet senior leadership. Make a list of people that you absolutely want to meet, and make a schedule of sessions and events that you absolutely want to attend. Contact people at least 2–4 weeks in advance and request informational meetings to take place over coffee. Bring with you:
  - business cards
  - pens
  - small notebook.

If I was on the job market, I would also bring copies of my résumé or CV because you never know if someone wants to look at it. But I would also bring those electronically on a flash drive or in a format on my cell phone or tablet that would allow me to quickly either show or deliver this document to another party.

- Use the conference app and social media: Most major conferences now have an app you can download onto your smart phone or tablet that allows you to organize the sessions and events you want to attend. This is such a great tool which in many cases is either not used at all or underused. The app can assist you in determining and sticking to your schedule. In addition, most conferences have a social media presence, and those that don't now, soon will. Conferences have Facebook pages and LinkedIn groups which you should join early on. And if they don't exist, why not create one yourself?

Learn the conference Twitter hashtag and handle for the sponsoring organization and follow the tweets from the conference. I can't tell you how invaluable this tactic has become just in the last couple of years for me as I attend more and more conferences. I follow the hashtag of the conference to learn about exciting sessions that I might not have realized were going to be so important, and to connect with people. At one recent conference, I was sitting in a session that I was finding to be a little boring and not the best use of my time. I whipped out my phone and took a look at the Twitter feed for the meeting and discovered numerous tweets from multiple attendees lauding the value they were receiving from another session taking place at the same time. I immediately made a quiet and quick beeline for the door and ran down the hall to this other session. And as it turns out, this other presentation was indeed much more interesting, and focused on more relevant information for my immediate needs than the one I snuck out of.

TIP: Follow the conference hashtag on Twitter to gain clandestine information about events, activities, and leaders. And don't hesitate to tweet as well, with the meeting hashtag and the association handle.

You can also use the Twitter feed to connect directly with other attendees. At another conference, I noticed someone tweeting a lot and I tried emailing her but she didn't respond. However, when I sent her a tweet with the conference hashtag, she replied right away and accepted my invitation to meet for dinner. Yes, it was in a public forum, but in the end it worked out great for me – I had a private meal with this person and solidified our partnership.

As I discuss in greater detail in Chapter 8 on Social Media, not only should you follow the tweets but you should consider tweeting yourself. This is an excellent avenue to establishing yourself as a thought leader in your scientific and engineering community. As people see you tweeting about the conference, they will begin to appreciate that you are someone they should take notice of, and follow. And this then leads to more people "following" you on Twitter.

- When you get there: As soon as I get to the conference and register, I immediately take out a bunch of my business cards and put them in my name tag holder. I then attach a pen to the side of the holder for easy access. I put my mini notebook in my hand so I also have that available. Having gone to so many conferences and I have so many name badge holders, I have gotten in the habit of keeping and reusing the really high quality ones. For example, I have a cloth name badge holder which has separate pockets for my business cards as well as business cards I gather, a pen holder and a zipper compartment for my notebook. So I have everything I need right at the ready when an opportunity presents itself.

- Scope out the venue, so you can get to know where things are happening and how to get there. This will aid you in adjusting your schedule as needed. For example, if Session A takes place on one end of a convention center and Session B takes place immediately after on the other side of the building (or worse yet, in a different building altogether), then you may have to modify your schedule to only attend one or the other. I also note possible places that are conducive to conversations to take people with whom I might meet. An exhibit hall during a coffee break can be a difficult place to have a useful conversation, so it's usually helpful to have identified some places with seats outside of the main, high traffic areas. Whether you have an appointment with someone that you have made in advance, or endeavor to have an impromptu discussion with someone you just met, being able to suggest a quiet location for the meeting highlights your aptitude to think ahead and your attention to detail. I have actually had colleagues take note of my ability to have planned locations for appointments that allow some privacy. And of course, high-impact networking at conferences demands that you and the other party can actually hear each other and have a solid conversation.

- Be flexible: Despite the fact that your plan involves a schedule, be flexible and open to new opportunities while you are there. Seek to attend presentations on subjects

> **TIP:** At conferences, there is no reason for you to ever eat a meal alone or with your current group members.

unfamiliar to you. Keep an eye out for impromptu or unscheduled events on the message boards (mobile or otherwise) that might provide you value.

- Take advantage of meals, breaks and evening events. At the conference center, during lunchtime I often see people eating by themselves or sitting against the wall reading their email on their laptop. But this is a great time for networking. So I try to introduce myself to people during lunchtime as much as I can.

- Take advantage of the trade show: Don't hesitate to go to the tradeshow and converse with the salespeople – this is an excellent opportunity to practice your brand statement and become comfortable with interacting with strangers. You never know what information you may acquire: You could learn of a new job or a new paper or a new field that you might find tantalizing. Gather information on the organizations exhibiting and exchange business cards. The trade show offers a seasoned scientist and engineer another priceless opportunity: To observe salespeople in their element. Companies and organizations that exhibit at a trade show are there for one purpose: To promote their brand and to convince you to "buy" their wares. This could come in the form of a company literally trying to sell a product such as a new type of microscope, a university trying to recruit graduate students or postdocs, or an organization trying to attract new hires, like at a career fair. In all cases, the organization behind the booth is there to convince you that they would be a good partner for you in your career in some way. So since the trade show is all about selling, the professionals that usually represent the organizations at the booths are seasoned salespeople, from whom you can gain powerful insight about how to market your expertise.

TIP: Trade shows serve as excellent networking nodes at conferences.

In addition, the trade show is valuable because it will give you an idea of who is recruiting and who is interested in your field of study. I found this recently at a seismology conference. I was amazed at the vendors who were exhibiting – there were companies I had never heard of. And even though they were there primarily to try to sell their instrumentation or software packages to the scientists and engineers in attendance, I realized that at some point in the product development process, the companies had to have engaged seismologists, either as employees or through consulting, to help them design, manufacture, tweak, troubleshoot, and sell the objects, as well as train others to use them. So what I really discovered from the trade show was a hidden job market for seismologists just by chatting with vendors trying to peddle widgets.

The trade show also naturally attracts other conference participants and thus provides terrific networking ROI. You will

TIP: Use your time at the trade show to network, rather than trick or treat for the free goodies.

undoubtedly discover that by just strolling the aisles, you will find people with whom you'd like to converse.

And finally, a note about all of the "free stuff" available at trade shows: Salespeople bring the branded pads, pens, squishy toys, and thumb drives to their booths specifically to catch your eye and entice you to come over and learn more about their enterprise. You can certainly partake of these trinkets. But do remember that just as a networking event is not about the food, a trade show is not about the free stuff. So don't get greedy or grabby with the pens and notebooks. Maintain your decorum and use the time more wisely to discuss potential avenues for alliances.

- Take advantage of the poster sessions: At many conferences, the poster sessions and the companion networking mixers that are organized in the poster hall offer very rich networking opportunities. You can review what posters might be of interest to you beforehand via the program and the conference app, but also give yourself the time to walk through the poster farm to see if anything new attracts you. I have found that by building in time into my conference schedule specifically to browse the posters I have achieved surprising networking ROI. The poster presenters are there for a reason – they want to speak about their research expertise so they are at the ready to initiate contact. But the other attendees, those who are browsing, are also interested in networking. So if you find yourself at a poster with another party, whether the author is there or not, use the subject of the poster as a way to strike up a conversation with this stranger.
- Introduce yourself and follow up: It bears repeating that you are at the conference to meet people, gain information about your science or engineering field, learn skills, and expand your career horizons. So when you attend sessions, come a little early and introduce yourself to the people sitting next to you before the talk begins. After the speech, go up and introduce yourself to the speaker. Ask everyone you meet for their business card or contact information, give them yours (see below), and follow up with them. Send thank you emails or notes to those who spend time with you and give you leads. At the lunches, in the hallways, near the posters, and by the cybercafé, introduce yourself and get to know your colleagues. Stay in touch.
- Act and dress professionally: Bring and offer business cards that list your current position (for example, "Candidate, PhD, Geosciences") and contact information. Use a professional email address (i.e., relegate sexyscientist@gmail.com to your midnight Myst trysts). Every STEM field has its own culture and social norms which dictate what the appropriate garb is for a meeting. You should get to know what your compatriots consider to be professional dress (be mindful of what they wear to conferences or, if attending for the first time, ask your mentor), and then bump it up a notch, especially if you are looking for employment in the near future. You are attending as a professional, which simply means you are serious about your

craft, and your wardrobe should reflect that. If you are speaking or presenting a poster, it's generally a good idea to wear a suit, or at least a nice pair of pressed chinos, a button down shirt, and possibly a jacket. And even as an attendee, you should wear clothes that are a little more polished than what you wear to the lab or in the field. At dinners and other events involving alcohol, monitor your intake – this is not spring break.

- Volunteer: As I touched on above, for students and emerging scientists and engineers, volunteering at a conference is your ticket to achieving more of your conference (and career) goals than you thought possible. And quite frankly, very few people take advantage of this opportunity. Volunteering at a conference establishes you as a professional and a hard worker, allows others to observe your dedication to your craft and the association, gives you easy access to networking opportunities, and opens doors to leadership and other volunteer experiences as well. Imagine if you were volunteering at the registration desk or in a session room. You are perceived as the authority because you are working at the conference. So not only will people automatically come up to you to ask your help (even if it is to inquire the location of the restroom), but you will have an immediate and very natural way to strike up a conversation. "Here is your nametag, Mr. Gore. I am so glad you can attend the American Geophysical Union Fall Meeting, and I hope you enjoy the climate change sessions. By the way, I am a student at Z University and am really interested in science policy. Is it possible I could take a few minutes of your time sometime during the conference to seek your advice about career options?" Here are just a few of the potential volunteer assignments you could land at a conference:
  - Registration desk.
  - Help desk.
  - Press Room: Gives you access to all of the media, as well as the public information officers who are hoping to use their conference participation as a springboard to get publicity about their own institutions. See below for more information.
  - Career/Job Center: Gives you unparalleled access to, and a reason to chat with, employers as well as those looking for a job.
  - Room monitor: You assist the speaker with anything they might need and even introduce them to the audience. They appreciate your help and remember you – what a fabulous way to connect with a leader in your field!
  - Blog: Volunteer to write a daily blog of the conference for the association's website. It will sharpen your writing skills, give you a chance to interview new contacts, and magnify your brand to decision-makers.

Furthermore, many associations have special events at their conferences to recognize volunteers, such as receptions. They might even have designated a lounge for volunteers to grab lunch. These networking nodes enable you to interact with other volunteers and meet the society leadership, thus increasing your networking ROI at the conference.

- Hang out in (or near) the Press Room: This is of great value if you have an interest in a career in science communications. You'll get to meet reporters and public information officers (PIOs) who represent various industry-related publications and institutes. You'll be able to talk shop with the journalists (just realize that they are there to do a job so if they are in the middle of something, recognize they could be on deadline). You will also be able to learn about the latest newsmakers in your field and find out about research trends. At the smaller conferences, the press rooms are more accessible to non-journalists and are more open to an early-career scientist or engineer stopping by just to chat with a reporter about their career trajectory. The larger meetings tend to have strict rules relating to who gets access to the press room, but it is still possible for you to gain entrance into the room. Start by going up to the room monitor and introducing yourself and stating your interest or goal for accessing the press room. Aside from getting career info and guidance from the writers, editors, and PIOs who attend, you also make yourself known to the PIO of the organization who is running the conference. This can be a very strategic and important move to make because the more that person knows about your research the more they can promote it to the reporters in attendance or throughout the year.
- Even if you can't attend ... : There is still networking value of which you can take advantage remotely. Review the program, see who is speaking or attending, notice what organizations and companies are exhibiting.
  - Send emails to particular presenters: "I saw that you are speaking at the 2014 Conference in Taipei. Unfortunately I won't be able to attend, but I am very interested in your research on X and Y and was really looking forward to hearing you speak about it. I would love to chat with you after you return. Would it be possible to make a short phone appointment in the weeks after the conference? I would appreciate the opportunity to explore a potential collaboration ... "
  - Send emails to exhibitors: Almost always when an exhibitor list is published, which you can access via the web, there is a contact person associated with the listing. If there is a company or organization that will be participating in the trade show with which you are interested in connecting, don't hesitate to reach out and send an email requesting a phone appointment. Remember, they are there to market themselves to potential clients or collaborators, so they will be very eager to speak with you.

- ○ Follow the Twitter feed. More and more, I utilize the Twitter feed as a window into a conference that I am unable to attend. The feed provides data, articles, pictures, and even video from the meeting and specifies the trends that emerged. By following the feed via the conference hashtag and the society's Twitter handle, I can also scope out possible contacts and "follow" them.
- ○ Access any online activity: Many conferences post videos of their keynote talks, and you can read the daily blog to find out what significant activities took place. The online reports will give you ideas and inspiration of people with whom you might want to network or organizations that might offer potential for career exploration and advancement.
- Follow up: After you make contact at the conference, you must follow up. Remember, it's not networking if you don't stay in touch; it is simply a waste of time and paper (on which the business cards you collected and distributed are printed). Follow up emails from making someone's acquaintance at a conference can be as simple as the following:

Subject: ESA 2013 – Nice to meet you

Dear Dr. Annette,

It was a pleasure meeting you at the Ecological Society of America (ESA) 2013 conference in Minneapolis last week. I appreciated the opportunity to briefly chat with you and X and Y (after your talk, while we were waiting for Dr. Z's sessions to begin, in the exhibit hall, at lunch in the convention center café, etc.). As you recall, I am finishing up my postdoc in aquatic ecology at the University of Arizona and I am especially interested in freshwater megafish. I found our discussion very interesting and I can see that there is definitely some synergy between our research areas. I would love to continue our conversation and explore the possibility of us collaborating. Would it be possible to arrange a short follow up phone/Skype appointment in the coming weeks? Please let me know what dates and times works best for you. I am based in Tucson and am on Pacific Standard Time.

   Again, it was great meeting you. I invite you to visit my LinkedIn profile at www.linkedin.com/in/alainaglevine and of course if you would like to see my CV, just let me know. Thanks for your interest and I look forward to speaking with you further and to perhaps working together in the future.

Best regards,
Alaina G. Levine

I usually send my first round of emails within a week or two after the conference. If I don't hear back from people that I have identified I really want to speak with soon, I often will send another email about a month later:

> Subject: ESA 2013 – following up
>
> Dear Dr. Annette,
>
> I don't know if you received my email from a few weeks ago (please see below), but I wanted to follow up with you about making an appointment to speak with you. I really enjoyed conversing with you at the ESA 2013 conference, and I think there might be an opportunity to collaborate. Would it be possible to set up a phone/Skype appointment in the next month? Please let me know what works best for your schedule.
>
> Thanks again,
> Alaina

TIP: One way of organizing your business cards is to catalog them by the conference that you attended. If you keep your name badge, you can use that to hold the cards you collected.

If you get busy and don't get a chance to follow up with a prioritized list of new contacts from the conference, don't think that all hope is lost and you can't and shouldn't contact them at a later date. Once you have their contact data, you can always email them. I recently went to a conference and didn't follow up with one particular person I met until four months later. It's never too late to contact someone!

A few years ago, I attended a scientific conference for which a high school student had written to the organization in advance and begged to volunteer. He was granted that chance, and the organization made a big, public deal that this kid had risen above his peers to pursue this unique and fruitful opportunity. It

TIP: It is never too late to reach out to someone you met (or wanted to meet) at a conference.

demonstrated his character, work ethic, and incredibly ambitious nature to want to be with and work with scientists in the field. The organizers of the conference spoke publicly about his skills, and I am certain he received offers to work in labs and apply for special scholarships just from that one experience.

You are not a kid. But that doesn't mean you can't and won't make a big splash at your next conference and dazzle those around you with your purposeful actions, enthusiasm, professionalism, and excitement surrounding your desire to hone your craft. So before you attend your next conference, contact the local organizing committee and inform them that you would like to volunteer at the event. Chances are they will take you up on this. Volunteering, coupled with a strongly executed strategic plan to achieve your conference mission, will ensure that you truly get the most out of your conference participation. You may find that the multitude of valuable career advancement doors that will open because of your active participation are too numerous to count. And such is life in a multiverse.

Articles

You already know that ideas and inspiration can come from anywhere, and that a diversity of ideas and sources of ideas are absolutely needed to succeed in science and engineering. So as you look for people with whom to connect, you should absolutely be reading articles to find potential collaborators. But what most scientists and engineers do is ONLY read the journals in their field (or more specifically their subfield) and then take no other action. They don't think of the article itself as a doorway to networking. But it is, and so much more.

For every paper written in a journal, or for any news article in which a scholar is quoted, very few people take the time to send that author or source a quick email to comment about their contribution. But you should, because as I wrote above, people are flattered when others note their public works and then take the time to approach them and talk about it with them. This action is so appreciated, and it can help build a truly solid win-win partnership between you and the other person.

I know this from experience. Time and time again, I have written emails to authors of articles or to people who are quoted in the media to praise their work and thank them for their input. And I almost always ask for an informal conversation to discuss the topic at greater length. They are happy to do so – it is unusual to receive responses to something you write, except if it is from someone complaining about it.

Let me illustrate this point with a story. In 2003, while employed at the UA, part of my job responsibilities included public relations and marketing. So I took out a subscription to *PR Week*, which is one of the leading trade publications for the PR industry. I was reading it casually over lunch one day when I spotted a really creative article. The writer, Lauren Letellier, had posited that the Harry Potter books, which were at their height of fame at the time, were possibly giving journalists a bad reputation, because one of the characters was a journalist who made up stories, lied, and cheated. Her thesis was that these books were potentially convincing kids that journalists (and by extension others who work in the media) are not to be trusted, and the profession of journalism is one that relies on lies and profiteering as measures of success.

I thought this was such a clever idea for an article and that her opinion was important – after all, children were being raised on those tomes and their impressionable young minds and preconceived notions about communications and its place in society were at stake. Very heady issues, indeed. I immediately realized that I wanted to contact Lauren and discuss this issue further. And I grasped that by simply writing this piece, she was inviting like-minded individuals to connect with her. In other words, she was creating her own networking node and opportunity!

So now I was very excited about the prospect of interacting with her and I needed her contact info. At the end of the article it listed her name, title, and company – Lauren was a senior vice president with one of the world's most renowned PR firms. It didn't list her email address or phone number so I quickly got to work tracking this information down. I couldn't find it on the web – this was before Facebook and LinkedIn. But her employer was identified, so I scoped out its website and found the main phone number for the headquarters.

Soon I was on the phone speaking with an operator. I didn't expect to get this professional's email address, just her phone number, so I was delighted and surprised when I was easily supplied with both. I have tried this tactic many times since when I am eager to reach an employee of a company and don't have their contact information; I will often call the main switchboard. Different organizations have different rules about releasing information, and if the operator had indicated that her contact information could not be released, I probably would have asked to be directly connected to her line. Many companies will do just this, so as a tip, if you do call a firm and end up being switched over to the person immediately, have an introduction ready.

Ok so now I had her email address. I wrote a draft of a fairly short cold email (in Word), following the template I described in Chapter 5, polished it up and sent it out. Afterwards, I sat back in my chair and relaxed, as I felt I had done my good networking deed of the day. Within 24 hours, Lauren responded with a positive and appreciative email: She would be more than happy to speak with me. We arranged a date and time, I confirmed her phone number and then I went to work preparing my questions. I had only asked for about 20 minutes so I wrote down only a handful of questions for her. And because I keep track of my networks and conversations I have with people, I was able to find the notes I kept from this conversation that occurred on 2 September 2003. Among my discussion topics, I asked her the following:

- What is PR agency life like?
- What has been your most interesting project or client?
- What are the best things about working for an agency?
- What are some of the things that could be improved?
- What are key things that one should know if making the transition to an agency?
- What is the agency structure and hierarchy like in your firm? Is it different in every agency?

- How do I determine what position my skills and experience match at an agency?
- How long does one work on a specific client or client project?

Lauren was so accommodating and pleasant and friendly, that she remained on the phone for approximately one hour despite the fact that I had requested less time. She clearly enjoyed the conversation as much as I did and was very open about issues pertaining to career tracks and even billing hours. And one of the last questions I asked her was "who else would you recommend I speak with?" and she quickly gave me the names and contact information of three people, including two high level executives with her firm and a headhunter who specializes in recruiting public relations practitioners.

As we concluded our discussion, I mentioned that if she ever has any projects relating to science, the southwest United States, Arizona, education, or anything related, that she can feel free to call upon me, and if I don't know the answer, I will find the right person who does. Then I got her mailing address, thanked her again for her time, and bid her farewell.

It's hard to believe that the conversation took place over a decade ago. It benefited me in so many ways. It gave me inspiration to pursue my career in novel ways that I hadn't considered previously. It gave me a positive role model of a woman who had achieved her own career dreams in PR and communications. It gave me the chance to connect with a creative soul which is often a rarity in any business. In short, Lauren gave me ideas, inspiration, and insight, and it all unfolded because I read her article and took the initiative to contact her.

> **TIP:** Make it a habit to regularly read articles (in journals and non-technical publications) in your sought-after profession, industry and sector. Seek to contact two new people you have identified from reading (either as the author or the source who is quoted in the piece) each month.

And incidentally, as I was editing this book, I emailed Lauren to ask her permission to tell this tale. She responded quite positively, and I learned that in the time since we last spoke she had left the public relations profession and had become a writer and playwright. In fact, her comedic play, The Fiery Sword of Justice, was performed during the prestigious 2014 New York Fringe Festival. As she wrote to me in an email, "Your career has certainly taken an interesting turn. After a lifetime working in PR agencies, I'm attempting to do something similar."

I can't wait to pick up our networking where we left off and explore new, perhaps comedic, avenues for collaboration!

Imagine all of the networking you can do just from contacting people who write or are quoted in articles! The opportunities are seemingly endless! But what publications, websites, and blogs should you be reading? Organize your reading list around these themes:

- Your desired discipline: Journals, newsletters, and websites of the professional associations, blogs, Facebook pages, LinkedIn groups.
- Your desired sector: Higher education, non-profit, government, private companies.
- Your desired industry or profession: Aerospace, environmental science, journalism, oil and energy.
- Your desired geographic areas: Localized newspapers, magazines, websites, Facebook pages, LinkedIn groups.
- Your desired employer or collaborator: Newsletter, websites, blogs, annual reports of the organizations, companies and institutions you would like to work with or for.
- Broad-scope publications: *The New York Times*, *The Wall Street Journal*, cnn.com, bbc.com.

TIP: When someone's honor is recognized in a publication, they appreciate being congratulated by members of the readership. Take this opportunity to offer them plaudits for their success and connect with them for networking purposes. You are not "sucking up" – you are extending a proposal to craft a win-win partnership!

Don't limit yourself to publications or sections of publications that are related to STEM. In other words, don't just read the Science Times section of *The New York Times*. Read the whole paper – and concentrate on finding potential collaborators in novel sections, such as business, arts, the Sunday magazine, fashion, and so on.

The business sections of newspapers, as well as magazines that are devoted to business endeavors, are especially rich with potential collaborators. For example, if you just read the following publications, you probably would generate an extensive amount of networking leads:

- *Fast Company*
- *Entrepreneur*
- *Inc.*
- *Forbes*
- *Fortune*
- *Wired*
- *MIT Technology Review*
- *Harvard Business Review*.

In addition to the articles themselves, there are also various sections in publications that highlight certain milestones and achievements of professionals in that field or in that region. For example, as part of the business section of my local paper, *The Arizona Daily Star*, there is a segment entitled "Moving Up" where professionals can submit notices about new jobs, awards they have received, and other professional triumphs (this is of course a great self-promotion opportunity!). There is a similar section in *SPIE Professional*, which is the quarterly publication of SPIE. There is no reason not to contact people who are listed in this section and congratulate them on their victory. They will appreciate your well-wishes and you can immediately schedule a conversation to explore the potential to partner.

Regional Industrial Representatives: Regional Economic Development Organizations, Chambers of Commerce, Industry "Cluster" Associations

The local economic development community is yet another concealed cache for networking with decision-makers in myriad sectors, both in and beyond academia. Many scientists and engineers don't realize the value to be had from connecting with economic development professionals and industry trade groups that are regionally based. Even if your career aspiration is to stay completely in academia, there are plenty of wise reasons to gain a better understanding of the industrial landscape that feeds and nourishes academia and vice versa, particularly in a specific geographical region. These are just some of the questions that can be answered by creating a dialogue with economic development professionals:

- What are the major high-technology sectors in the region?
- What companies employ graduates and alumni?
- What companies support internships?
- What companies license technology from the university?
- What companies financially support university research, either through monetary gifts, in-kind donations of laboratory equipment, or through sponsorship of projects?
- What company leaders sit on university industry boards, and what university leaders sit on corporate and community economic development committees?
- What university leaders sit on regional (and global) industry boards and committees?
- Is there a Memorandum of Understanding (MOU) between certain companies or economic development organizations and the academic institution?
- Is there a strong entrepreneurial community?

Now of course the reason you would want to ask these (and many other questions) of the local economic development community is because you will learn of hidden career and project opportunities and network with the professionals who know about them and have the power and authority to give you access. But many scientists and engineers, when researching a location to which they might want to relocate, overlook these fantastic and profitable resources of information and networking and therefore miss out on the dividends they provide. So if you were to engage the economic development community in your region, you might just be surprised at what you would be able to gain.

Most municipalities in the United States and around the world have regionally-based economic development agencies or organizations whose charter it is to ensure the economic welfare of the region. They do this .8through many different types of outreach projects, all with the same basic objectives: To attract, retain, and grow high-growth industry, and to attract, retain, and grow talented, credentialed professionals who will launch new companies in these sectors, and serve as employees for companies that are already there. It's a cycle driven by money: The more high-growth industry we have in a location, the more high-credentialed talent we can attract, which means the more industry we can grow attract and retain. And with those high-growth companies and talented professionals come taxes and other economic benefits to the community. It's all about devising an ongoing, self-feeding, win-win relationship between a community, its residents, and its businesses.

> **TIP:** High-growth industries tend to be focused on technology, science, and engineering, which naturally attracts higher-educated professionals who fill higher-paying jobs.

The different organizations that are involved in economic and work-force development initiatives are on the pulse of what is happening locally in particular industries; they know who the leaders are; they know who is hiring and what skill sets they are looking for. In short, they can provide you with strategic information about people, companies, and opportunities that can lead to job offers. In fact, because they are heavily involved in the growth of industries, they often know about job openings (particularly a large number of openings, like when a company is about to pursue an expansion or an acquisition) before the general public. So it is extremely important for you to network with the economic development professionals in the regions in which you are interested in working so you can access these hidden opportunities and learn about strategic information which can assist you in your career planning and job hunt.

Furthermore, by connecting with the economic development organizations in the region you can get on their mailing list and be apprised when they have networking events, which often take place many times throughout the year. Depending on your location, these are

> **TIP:** Economic development professionals often have inside and non-promoted information about upcoming or current job opportunities, career resources, and hiring trends in their region. Get to know them!

often attended by executives in industry and academia as they are often seen as a must-attend event for networking. So if you know about these events you can attend them.

It is important to know the different types of organizations that exist to advance local economies. There are a number of layers of economic development offices and organizations within regions and each municipality has their own way of doing things. But below are the basic organizations and stakeholders you probably would find in most cities and large towns in the United States and elsewhere.

- Economic development divisions of local and state government: Within the city and county there is often a department of economic development. And most states in the US have some sort of department of commerce which also employs economic and workforce development professionals.
- Regional economic development organizations: This can be a government-run agency, a private non-profit, or a hybrid that feeds off both public and private investments. In Tucson, our regional economic development organization is called Tucson Regional Economic Opportunities, Inc., or TREO. On its website, TREO is described in this way:

> Tucson Regional Economic Opportunities, Inc. (TREO) was formed in July 2005 to serve as the lead economic development agency for the greater Tucson area and its surrounding regional partners. The primary goal of TREO is to facilitate export-based job and investment growth, in order to increase wealth and accelerate economic prosperity throughout Southern Arizona.
>
> This work demands a competitive, business-friendly environment that allows primary employers to flourish and succeed. Thus, a secondary role of TREO is to shape policy and mobilize resources to ensure the region is competitive. TREO engages in various efforts and partnerships focusing on demonstrating leadership to strengthen education, create a vibrant downtown and engage in infrastructure improvements.
>
> To serve a population approaching one million residents, TREO offers an integrated approach of programs and services that support the creation of new businesses, the expansion of existing businesses within the region, and the attraction of companies that offer high wage jobs.
>
> As the region's only true private/public partnership, TREO connects the private sector, governments, nonprofits and academia to provide leadership on competitive issues and a unified voice to the national and international business community. TREO is supported by a Board of Directors that represents both private sector businesses and public sector partners.[ii]

Economic development organizations are often membership-driven, and the members are usually companies and institutions in the area. These organizations conduct constant research to stay on top of the hiring and economic growth trends in the region, and engage in heavy public relations and promotional activities to ensure that this information gets to decision-makers, such as site selection consultants for companies that are considering locating to the area. What this means for you is this: They have data about career and job opportunities, they have access to the major players who can give you access to these opportunities, and they have events that you can attend so you can network with the decision-makers yourself. Their websites list the major employees and industries. Their reports go into detail about how the economy is operating and organized.

The economic development organizations often host events that are considered high priority for captains of industry to attend. In fact, in Tucson, one of the biggest business affairs is the annual TREO Luncheon which attracts hundreds of people from myriad industries and sectors. It is considered "the" event of the year by many professionals and the networking opportunities are huge and precious. A few years ago, I attended the TREO lunch and reconnected with a former colleague while applying lipstick in the ladies room. It directly led to a consulting gig.

- Regional chamber of commerce: "A chamber of commerce (also referred to in some circles as a board of trade) is a form of business network, e.g., a local organization of businesses whose goal is to further the interests of businesses. Business owners in towns and cities form these local societies to advocate on behalf of the business community. Local businesses are members, and they elect a board of directors or executive council to set policy for the chamber. The board or council then hires a President, CEO or Executive Director, plus staffing appropriate to size, to run the organization."[iii]

Since the chamber exists to promote business in a region, it welcomes members of the community to serve on its boards and committees, and to attend its sponsored activities.

TIP: If you have a particular region in mind, connect with their economic development teams in advance of moving there.

Years ago, I learned of a Tucson Chamber of Commerce committee devoted to business–education partnerships. I joined and found myself networking with the principals, superintendents, and other educational leaders. These compatriots soon became collaborators as we undertook the design and implementation of outreach programs that benefited K-12 STEM education.

But keep in mind – you don't have to only network with your local chamber! A colleague of mine knew she was going to be moving to

Tucson but wanted to get a better comprehension of the economic and industrial landscape of the city and region before she finalized her plans. She called the Tucson Metropolitan Chamber of Commerce and spoke with the receptionist about her interests, who connected her to someone in the Chamber's Workforce development program. My colleague made an appointment with the Chamber representative to meet with him face-to-face the following week when she was visiting the city on a house-hunting trip. That meeting resulted in her being introduced to a number of leaders in her field and yielded a job offer in Tucson even before she got to town. So even if you think you are going to move to a new place, or are just researching it, you can reach out to the regional economic development organizations and ask for a meeting, even if it is by Skype or phone.

- Industry clusters: "A business cluster is a geographic concentration of interconnected businesses, suppliers, and associated institutions in a particular field. Clusters are considered to increase the productivity with which companies can compete, nationally and globally."[iv] Clusters are very important in terms of economic development – when you have a critical mass of companies of a particular industry, that helps to attract other companies and suppliers, as well as talent to be employed by those cluster companies. It also fosters an entrepreneurial ecosystem that allows for the development of new companies in the region.

And the most wealth-producing clusters are those focused on high technology.

You probably have heard of the most famous high-tech cluster: Silicon Valley. But there are other well-known clusters. New Jersey has a major cluster of pharmaceutical and biotechnology companies, as do Boston and San Diego. Los Angeles has a cluster of aerospace companies, as do Boulder and Seattle. Of course, clusters don't necessarily have to focus on high technology, as a region could have a cluster of companies in hospitality, service, or manufacturing. For your purposes, the high-technology clusters will be the most advantageous to know.

Because a business cluster provides a boon to the companies in the area, representatives of that industry work very hard to keep the clusters healthy and growing, by seeking to continuously attract new talent, outside investment, and the development of new firms. One of the slickest ways to achieve this umbrella objective is to bind together in a industry-focused trade organization. The industry leaders essentially get together and form a non-profit society, with elected leaders, members, and events, with the purpose of ensuring the cluster's prosperity. They often have job boards and listservs and LinkedIn groups. They provide newcomers to an industry and/or a region a link to the hiring managers and trends that can truly enhance a networking experience.

Often, high-tech clusters are buoyed by nearby universities. For example, the UA College of Optical Sciences is world-renowned and many of its alumni remained in Tucson upon graduation and formed optics companies. There are now so many optics companies that Southern Arizona has a bone fide optics cluster and the area is often referred to as "Optics Valley." The cluster has a non-profit organization entitled the Arizona Optics Industry Association (AOIA), which elects board members, has meetings, and communicates industry information through various channels. It also serves to advocate on behalf of optics companies and professionals at local, state, and even national fora. If you were interested in a career in the optics industry in Tucson, I would recommend going first to the website of AOIA, reviewing what the major companies are and who their leaders are, and then sending emails to some of AOIA board members explaining that you are interested in moving to Tucson (or if you are already here, that you are interested in transitioning into the optics industry) and would like to make a phone/Skype/coffee appointment to learn more about AOIA and the local industry, and how you can become an engaged member of AOIA.

The bottom line is this: Economic development agencies want to speak with you! Scientists and engineers, as mentioned above, are the value creators of society. STEM professionals, especially those with advanced degrees, are seen as extremely desirable and even required for a region to economically advance – you help grow companies, advance technologies, and create new industry through your entrepreneurial efforts. Economic development pros know that when high-tech companies consider relocating or launching a satellite office in a new area, they want to ensure they have a steady supply of educated workforce. This is what attracts and keeps high-tech companies in the area, and it also spurs the growth of new enterprises regionally. So to economic development professionals, whose goal it is to sustain high-wage jobs and high-tech companies (because they pay more) to an area, scientists and engineers are the lynchpin. They covet you. They want to hear from you. They want you to move to their city, start a family there, and stay there because your presence will significantly contribute to the region's economic affluence and influence.

Just as I have written above about taking full advantage of your membership in your professional society to network and promote your brand by attending events, volunteering, writing in the newsletter, and so on, so too can you pursue these activities within economic development organizations. In fact they hunger for outside participation. A few years ago, while I was working for the UA, I began reaching out to the Arizona Technology Council, a state-wide organization that advocates for tech-based companies. As written on its website:

The Arizona Technology Council is the driving force behind making our state the fastest growing technology hub in the nation; connecting and empowering Arizona's technology community. As Arizona's premier trade association for science and technology companies, the Council is recognized as having a diverse professional business community. The Arizona Technology Council offers numerous events, educational forums and business conferences that bring together leaders, managers, employees and visionaries to make an impact on the technology industry. Council members work toward furthering the advancement of technology in Arizona through leadership, education, legislation and social action. These interactions contribute to the Council's culture of growing member businesses and transforming technology in Arizona.

I could easily see that there was a synergy in their goals and the objectives of the program I was overseeing at the UA. Since the UA was a member, I was able to volunteer to serve on a committee. But I didn't just pick any committee – I volunteered for three years on the statewide awards committee. For the minimal time it took to participate, literally a few hours over the course of each summer, I was able to gain access to tech leaders with whom I probably would never have spoken. Furthermore, as I noted above, by participating on the committee and doing the work associated with it, I was branding myself as a hard-worker, an energetic and enthusiastic team member, and establishing myself as a "known quantity." The group knew I added value. So after two years when the chair stepped down, I eagerly volunteered to co-chair the team. Now I was front and center. And yes, I was able to gain a lot from the experience, including more insight into team dynamics, event management issues, and even office politics. And then, as I have stated so often in this book, I leveraged this opportunity to get another one with the same organization and started writing for their technology magazine.

The final piece of any regional economic development puzzle is the community of entrepreneurs. In fact, one could argue that the entrepreneurs make up the core of any economic development agenda, as they will take the city's industry into the future. When researching a particular municipality, conduct a Google search to determine how the entrepreneurial community is characterized, what resources they have available, if they are "incorporated" (i.e., they have an organization to further their aims) and what kinds of activities they sponsor. Look into the region's "Maker" groups, as well as websites or physical communities devoted to start-ups. Even if you are not an entrepreneur yourself or are not interested in working for or launching a start-up, there is still much that can be gained by connecting with the new venture community of a geographic region, for both you and the community. In particular, look for/at:

- Conducting a Google search using terms like "start up," "entrepreneurial/entrepreneurship," "maker," "venture capital," "angel" (see below) and the name of your city or town. You'll find out about articles, events, and organizations that support this community.

- The local college's or university's business school: Most larger business colleges now have some sort of entrepreneurial element, be it a formal degree in the subject or a center devoted to it. Business schools recognize the importance of teaching entrepreneurship or incorporating it into the curriculum in some way, and that often means that local entrepreneurs, venture capitalists, attorneys, and other professionals who are adept at entrepreneurship interact with students, via mentoring programs. This invariably leads to events that the business school holds, such as business plan competitions or networking receptions. Leverage the entrepreneurship program at the regional academic institution as its own networking node – get on their listserv, like them on Facebook, follow them on Twitter, and attend their gatherings, many of which are open to the public. You'll meet local entrepreneurs, as well as other business and technology leaders that support these types of enterprises, get a feel for the entrepreneurial and tech climate in the area, and make connections that can help you land jobs and other opportunities. Additionally, most business schools have entrepreneurship clubs that are always looking for speakers to discuss technology and business, or that hold meetings open to the public. Find out about these through basic Google searches and more extensive searches of the university's online club directory, offer to be of assistance and even be a speaker, and ask if you can attend some of their events. They'll be thrilled to have you! I myself taught a class on entrepreneurship for scientists and engineers for five years at the UA, and was always looking for guest lecturers, and professionals with whom my students could network. And in many cases, I had guest speakers address my class who did not work for a start-up, but still had valuable information to share about technology, business, and innovation. I also hosted a large public event at the end of each semester where my students would present the early-stage business plans they had worked on for the semester. Audience members included venture capitalists, entrepreneurs, tech/business leaders, economic development professionals, and university officials and professors. For students and attendees alike, the gala is networking gold.
- The local university's tech incubator and technology transfer office: Many major academic institutions have tech incubators which are meant to help faculty and students spin off technology developed within the university into companies. The new ventures incubate there, and are assisted by a team of well-qualified mentors, attorneys, venture capitalists, and other investors, and even accountants and public relations practitioners. They can attend skill-building workshops and networking affairs. You can get on the mailing list for

your local incubator and become a part of this important community, and you'll find out about job opportunities within these companies and their partners. Likewise, the college's technology transfer office, whose mission is to commercialize innovations and discoveries made at the school, is also well-connected with the entrepreneurial community. They may have public activities and listservs that you can join to learn more and find professionals with whom you can efficiently network.

- Entrepreneurial or start-up conferences, which may include start-up bootcamps teaching business skills needed for successful running new ventures, and pitch fests/slams where entrepreneurs get a chance to pitch their ideas to the community for potential investment. These offer extensive high-impact networking opportunities and give you the chance to learn the major players in different industries (especially high-tech) in the city.
- The regional "angel" group: In entrepreneurial parlance, "angels" are high-net worth individuals who invest in start-ups. They often form regionally-based clubs and have regular meetings at which would-be entrepreneurs pitch their start-up ideas to the members. The angels decide independently if they want to invest in a company; the group itself just serves as a networking node to bring together potential investors and entrepreneurs. Some angel groups require their members to write checks, others make it optional. Angels often participate in other local events and activities that promote entrepreneurship, so they are on the pulse of what's happening in their start-up community. You can search for your local angel organization by Googling "angels" and the name of your city.
- Meet-ups: There are lots of informal meetings, conferences, events, and even competitions that you can find on the website meetup.com.

Alumni Associations

Alumni Associations are fabulous, fabulous resources for networking which very few scientists and engineers think to exploit. Business majors know to join their alma mater's alumni associations and go to alumni association events and contact other alumni, but science and engineering majors rarely think to do this. But alumni associations are gold mines. And think of this – you probably have more than one alumni association! If you have an advanced degree from another institution, then you have at least two alma maters and for those of you who have a PhD from a third university, that's three universities whose alumni associations you can rightfully take advantage of.

A word about why alumni associations exist: Their only purpose is to provide an avenue to keep alumni engaged with their institution in order to generate development opportunities, that is donations, publicity, and new student enrollments. An engaged alum is a happy alum, and a happy alum is one who is more willing to donate to their alma mater, publicize their alma mater's many successes, and even enroll their child in the school. Alumni associations are eager to create many channels for ALL of their alumni to feel connected and engaged so the flow of support never stymies.

> TIP: Seek to attend alumni events in any city to which you are traveling.

Alumni association leaders have thought this through very, very carefully and believe me, the smart and most successful alumni associations, the ones whose institutions have the largest endowments, know that they must start this engagement process early. You become an asset not on the day you graduate, but on the day you enroll at the university. Do you see the parallels between what alumni associations are trying to achieve and what networking tries to achieve? Alumni associations want to create long-term partnerships with alumni that are meaningful and provide a win-win prospect for both parties. So how do they do this? They create and implement lots and lots of engagement activities that keep alumni constantly connected to the alma mater. The magazine, homecoming events, region-based chapters, and services like alumni travel programs and even career advancement programs are all paths to accomplish this. Anything they can do to keep you connected allows them the opportunity to cultivate a long-term relationship with you where you feel you are getting something in return and therefore are interested in giving something back, either in the form of money or mentorship to students or even recruiting students to work at your company.

Alumni association tactics:

- Join your alumni association! It is absolutely worth the investment.
- Start with the directory: If alumni are in the alumni directory that means they WANT to be contacted. So search the directory to find people with whom you would like to connect. And if you are simply exploring it, don't just look for professionals who graduated with your same degree in your same year. Diversify —look for people in different cities, companies, industries, and disciplines. Furthermore, add yourself in the directory too! Keep your contact information up-to-date, because you want to be found just as much as you want to find others. And of course you want to receive all of the association's announcements.
- Pore over the alumni magazine and other publications: These resources are incredibly valuable. You can learn about new programs at the university to which you could possibly contribute, and of

course, you will discover information about alumni activities and successes. If there is an article about an alumnus working for the United Nations and that is your dream, you have an immediate connection with this person. Contact him, and network.

- Promote yourself: If you have done something really spectacular in your career or overcome some adversity to persevere, or are just working on a project that is absolutely fascinating, consider contacting the editor of the alumni magazine and pitching an article to be written about you. You can also contact the writers of the magazine itself and let them do the pitching for you to their editor. Why shouldn't you get some great promotion about the work you do? And even if the editor or the writer passes on a full story about you right now, you can submit your victory to the "class notes" section which usually appears in the back of every alumni magazine.

- Volunteer for committees: Most alumni associations have a major council, which one usually has to be nominated to sit on. However, there are generally several different committees that engage alumni which do not require a formal nomination to join, and their leaders would welcome your contribution as a volunteer. You'll welcome the networking and self-promotional opportunities they will provide.

- Attend local alumni events: If you live in or near a major metropolitan area in the US, chances are your alumni association has a chapter. You can contact the leader of that chapter and engage him/her the same way as someone from your professional society. Ask to take them out for a cup of coffee and learn more about the chapter. Get on their mailing lists and start attending their events.

- Attend local events elsewhere: Whenever I travel, I always check to see if there is a local chapter of my alumni association in that municipality. I will often contact the president to arrange to meet or see if there is an alumni event happening at the same time while I am in town. This goes for when I travel to another city for a conference too. This tactic really helps to diversify my networks. Recently, I was heading to Chicago for a speaking engagement. I am on the listserv for the UA Alumni Association, as well as their LinkedIn group, and got a notice that the Chicago section of the alumni association was having a big event to welcome the new president of the association. It happened that the mixer was going to take place the night I arrived in the Windy City. It was free to attend, so I RSVPed, and went straight from the airport to the hotel where the event was taking place. There were a great many alumni at the event, but not so many that I felt swallowed up by the affair. In fact, I ended up meeting some very interesting people with whom I otherwise would not have had a chance to interact. I also had the opportunity to reconnect with the new president, whom I had known from a previous job, and introduce myself to other key UA representatives. All and all, the networking ROI was high. In fact, some of the leaders said they were impressed that I came to the event even though I live in Tucson.

- Consider joining other alumni societies associated with your alma mater: Your college, your department, and even your major might have its own alumni board, even if it is a very informal group. Your college probably has a board of directors of prominent alumni or industry leaders. See if you can contact them or better yet join it. Similarly, there may be alumni councils associated with different student clubs, such as the student newspaper and of course, the Greek System. Many campus organizations are affiliated with a national mothership, and you can leverage your student membership to become an engaged member of the larger group.
- Take advantage of your alumni association's career services: More and more associations are providing career services and resources for alumni, especially with the downturn in the economy. You can often leverage your membership to learn about job opportunities, gain or polish your soft skills, have your résumé critiqued, and of course, network with fellow alumni and alumni association staff who know where the hidden career opportunities lie.
- Go on alumni travel! It may seem like a silly notion, but alumni travel delivers high-impact networking ROI. Alumni associations sponsor study tours that attract other alumni, often in very small groups. Thus not only do you go on a fascinating voyage, often coupled with lectures from faculty, but you also get to bond with a manageable cluster of like-minded folks for at least a week. Consider yourself networked!

Regional Philanthropic Organizations

If you are interested in a cause, be it animal rights, women's issues, combating poverty or mentoring at-risk youth, the concept of volunteering in your community is probably not foreign. In fact, you may already instinctively contribute to various philanthropic endeavors in your region, either through gifts or by taking time out of your week to volunteer. Many people who participate in charitable engagements attest that they get more than they give. And in fact, one of the benefits that you can gain while supporting your community is finding people with whom to network.

As you support philanthropy, you'll notice like-minded individuals working alongside you. These are potential professional partners. You will find these people from volunteering and from pursuing leadership roles in charity organizations, such as the YWCA/YMCA, United Way, and various others. And since many cities have volunteer centers that help coordinate and promote volunteer opportunities, you can easily take the philanthropic pulse of a community by perusing the volunteer

> TIP: Since networking is meant to be a win-win partnership for both parties, you should not hesitate to network at any type of event, even one that centers around a charity or philanthropic cause. After all, when you attend, you are all gathered there to support the endeavor, so you ALL might as well get to know each other and your mutual interests relating to the philanthropy and beyond.

center's website to see what openings and charities are available. This is useful information to have if you are considering relocation to a new city, or even if you have been in your region for years. It is never too late to serve your community via its philanthropic enterprises, and it's never too early to research a region by engaging the charities that operate there.

Another option to consider when looking to network via philanthropic lines is attending events. Charity events, which range from \$500/plate balls to \$20/person "races for a cure," allow for extremely effective networking. The people who attend these are often leaders in industry and view their participation in these events as a strategic element to their company's public relations missions. And when they attend the affairs, they engage in high-impact networking. The charities know this and often promote their activities as potential networking goldmines. So if you attend, don't be a wallflower – go up and introduce yourself to others.

"Young Professionals" Societies

As I noted above, cities are eager to attract highly-skilled, particularly young, professionals because you help bolster high-tech industry, which in turn attracts other professionals to the community. The presence and influx of early-career scientists and engineers is seen as an indicator of economic health and wellbeing in a community. So economic development experts, academics, industry leaders, and even non-profit managers have, in recent years, united to strategically enhance their community's offerings for early-career professionals.

One of the outcomes of these vast partnerships is the creation of "young leadership" societies and alliances within communities. These groups function like a typical professional association, but their goal is singular: To attract and retain early-career leaders in the community and aid them in advancing in their vocations. For example, "Tucson Young Professionals (TYP) is a group of young business and community leaders focused on the promotion, attraction and retention of young professionals in Tucson. Together we create a voice to help impact the development and growth of Tucson."[v] TYP hosts regular events, participates in its own philanthropic enterprises, and even engages senior leadership in

Southern Arizona for dialogue and networking opportunity. Nowadays, you will find similar young leader societies in most metropolitan areas.

In addition to the stand-alone young leadership groups, charities have realized that the earlier in their career they engage someone in the charity's mission, the longer they will have them as an investor and stakeholder. So organizations like United Way have launched their own young leadership societies, which also offer a bounty of networking and self-promotional opportunities, while giving back to the community.

Religious and Political Affiliations

Networking with compatriots who share the same values as you is a privilege. You can often seek these types of unions under the umbrella of your religious and/or political affiliation. If you are "religious" (in any sense of the word, from extremely passive or reform to extremely active or orthodox), consider approaching and volunteering for your local church, synagogue, mosque or temple. And the same thing goes with your politics: Whether you are a democrat or a republican, a libertarian or a communist, a tea-partier or a coffee-partier, there is most likely a group nearby in which you can participate and network. And as you travel the world, look into popping in to the local house of worship or political headquarters for on-the-move networking ROI.

Regional STEM-related Activities

Beyond what academic institutions offer, there are usually numerous networking-centered events and activities designed to engage a STEM-enthusiast public. For example:

- The science or art museum evenings for adults: Whoever came up with this innovation deserves an award. Art and science museums around the world realized that their space was underutilized in the evenings (of course they were – they were often closed except for special events) and that they could attract new populations and thus recruit potential new members if they had evening parties just for adults. So they started having Adult Only evenings. Often coupled with or promoted as a themed-party, the adult-evenings at the museums are another great place to network locally. Once a month the Tucson Museum of Art has a Friday-night party, complete with music, dancing, food, and drinks, designed to be a social event for young professionals to network (although anyone of any age is welcome). People dress up and each month's gathering features a different theme, such as Carnivale (where they supplied masks) or disco (where they encouraged people to wear their finest 70s threads). And when you travel you should check out the museum schedule of the city in which you are in to see if there are any evening events going on. Take advantage of them. I was attending a conference in Dallas, and one of the conference's evening events happened to coincide in time and space with the city museum's evening reception for local professionals. As I finished networking at the conference, I migrated over to another area of the complex and joined hundreds of Dallas area adults enjoying the museum after hours and networking to their hearts' content.

- Science cafés: More and more of these are popping up in cities around the world. Some are organized by universities or other academic intuitions; some are organized by industry-related organizations; and some are flung together using social media sites like Meetup.com. They also range in what they offer – some feature a lecture on a topic (given in such a way that non-experts can enjoy it) with questions and answers either during or at the end, and some are more discussion-centered. No matter the topic, style, and mode of delivery, Science Cafés are a terrific way to network with likeminded nerds.

- Regional science fair: I was only a few years out of the university when I volunteered to serve on the committee that organized the regional science fair in Tucson, but given the responsibility with which I was tasked, you would have thought I had years of experience. As a result of volunteering at the science fair, I networked with science outreach professionals from the city as well as scientists and engineers who volunteered to judge the fair. I arranged for our US Congressman to give a keynote address at the fair and even wrote his speech. I accompanied our regional winners to the International Science and Engineering Fair where I further networked with science teachers, advocates, and leaders, professionals in business, media, and politics and policy. In fact, one evening I was at the science fair networking reception for "adults" when I sat down at the bar and had a conversation with someone who turned out to be the editor-in-chief of one of the most famous science magazines in the world. Thirteen years later, I ran into him at a conference for science writers and he remembered me!

- Science festivals: Just like science cafés, more and more regions are hosting science festivals which offer activities ranging from speeches and demonstrations given by STEM professionals, to receptions and parties for nerds. Go to these events and, better yet, volunteer to help organize them. You'll be appreciated, have access to amazing people with whom you might not otherwise be able to chat, and you'll get ideas of where you can take your career next.

Spontaneous Networking – On an Airplane, Train, Taxi

In addition to seeking out formal opportunities in which to network, there are many, many chances to spontaneously network with others as you go about the course of your day and business. I have already discussed a few examples of networking on an airplane, but here are some ideas for other spontaneous networking scenarios:

Offer to share a cab. When I attend a conference, I often share taxis with people from the hotel to the airport. In the reverse it is not necessarily safe, but when leaving from a hotel, where most of the conference attendees are staying and you both know the location to which you are going (like the airport or train station), it can actually prove quite fruitful. You can also employ this tactic to spur a networking opportunity if you are traveling on other business as well. I was giving a talk at Northwestern University and arranged for the hotel to call me a cab at some ungodly hour to make my flight in Chicago. When I checked out, they pointed to a women in the lobby and said she was also going to the airport – "could she share your cab?" the bellman inquired. Of course, I was delighted to do so. It turned out that the women was a science professor and ran a program for women leaders in STEM at her university. We clearly had a lot in common and our 30-minute trip to O'Hare seemed to go by very quickly. We ended with an agreement to follow up to discuss working together.

At large enough conferences, like those where 20 000 people arrive via airplane all on the same day, you can often spot who the conference attendees are by their clothing, stickers on their luggage, and a preponderance of poster-carrying tubes. In this instance, it is probably safe to offer to share a taxi with someone if you are both going to the conference hotel, although you should certainly be cautious about this – for example, as a woman, I might not necessarily do this with a strange man, whereas I could see myself offering to share a taxi to the conference hotel with another woman. In general, don't hesitate to strike up a conversation with folks in the taxi queue or even in line for the metro. For the last few years, I have spoken at the American Geophysical Union annual conference for in San Francisco. Over 22 000 people attend this meeting, and I see people carrying posters throughout my journey to the City by the Bay. And when I finally arrive at San Francisco International Airport (SFO), I always do so at what seems like the exact same time as throngs of geoscientists from all over the world. In fact, with so many planes landing at SFO in the same time window, the lines to buy metro tickets can be extremely long. But when you enter the queue, you can clearly determine who is there for AGU. So I have networked just by standing in line, and have even suggested to people that we share a taxi downtown, giving me plenty of time to learn about them, their research, and their objectives for a potential alliance.

Catch someone alone. Some of my richest networking ROI has been done when I saw someone, perhaps a senior member of a committee, a president of a university or company, or an editor, standing alone. Perhaps they were reading their email off to the side of a conference, walking to their engagement, or having a drink at the bar. Whatever the circumstance, when I saw that they were not otherwise engaged with another individual, I seized the opportunity to speak with them. I wasn't rude about it – in fact, I usually go up to the individual and say "Excuse me, Dr. Feynman. My name is Alaina Levine and I am a science writer. I have read all of your books and love how you communicate your passion for physics. Do you have a few minutes to chat?" The key is to be polite and recognize that there is a chance you are interrupting them. So be quick, be gracious, and ask if this is a good time to talk. If it is not, you can always ask for their contact information and ask if you can follow up with them to make a phone appointment.

Years ago, Steve Forbes, President of the Forbes Companies and Editor-in-Chief of *Forbes Magazine*, was on a book tour and happened to be making a stop at the UA. He was scheduled to give a public talk on the same evening I taught a class on entrepreneurship. I knew this would be an unmatched opportunity to hear a great speaker share his wisdom about business so I cancelled class and required my students to attend his talk instead.

I was very excited to hear him speak. I planned to arrive early to ensure I got a good seat in the auditorium, but as a result of some late afternoon foot traffic in the Student Union (where his engagement was taking place), I found myself running late and taking a back stairwell to make up for time. As I climbed the steps I looked ahead of me and to my surprise Steve Forbes was also ascending the staircase with a puzzled look on his face – it turns out he was lost.

I didn't think. I yelled out "Steve!" He turned and I ran over and extended my hand and introduced myself. He didn't stop walking so I just walked right along with him and matched his pace. I told him a little about myself and my class and that I had required my pupils to attend because I knew they would gain great value from hearing him speak. And then I told him that I write a column on marketing for the local business paper and "I'd love to interview you for it – would it be possible to make a phone appointment to do so?" He responded yes and I asked him for his business card. He didn't have one on him, so I gave him another one of mine and asked if he could please write down his email address on it so I could get in touch with him. He did so, all the while walking. I thanked him and then pointed him in the direction of the room. I emailed him about a week later and he responded politely and put me in touch with his assistant who scheduled the interview. And most importantly he said "stay in touch." To this day I still have his email address!

Don't be afraid to grab someone while they are alone or are en route somewhere. You don't have to make yourself a nuisance and if they are in a hurry you can simply and quickly introduce yourself, give them your card, thank them for something they have done or tell them something that you would like to speak with them about further, and of course ask if you can follow up. I have known numerous professionals who created powerful, fruitful, long-lasting alliances with people they caught while en route to the restroom. If the opportunity arises, seize it!

Other Places to Meet People

The colloquium speaker: Offer to give the colloquium speaker a ride to/from the airport, or better yet volunteer to be the host. You get private face time with him and can have a real conversation that can deliver real networking results. At the very least, ask if you can accompany the speaker and the committee to dinner, or even serve on the committee. A famous author was scheduled to speak in a little town in Colorado and my friend offered to drive her to the airport in Denver. What should have been a fairly easy trip turned into a scary seven-hour trek over mountain roads in the middle of a dreadful snow storm. My friend could not have been happier! Not only did she get a huge quantity of private time with this author but since they shared this unusual experience together, the two of them forged a bond. Furthermore the author will always be grateful to my friend for volunteering for this scary and courageous task.

When traveling to a new city: When I travel for a conference or a business meeting (and sometimes even on vacation), I will often look for opportunities to network away from home. This includes alumni chapter events in other cities, as I discussed above, and it also includes cold emails to people with whom I think I might have something in common. For example, if you knew you were going to a conference in Washington, DC, you could contact faculty or postdocs at one of the regional universities there. You could indicate why you were interested in chatting with them and ask if you could take them out for coffee while you are in town. You could also use the opportunity to check out the local calendar of events and see if there are any networking receptions or mixers that you could attend while visiting. Scope out the local business newspaper or the business section of the regional publication's calendar to see what is happening in the city you are going to. In some cases, it might do you some good to extend your trip by a day or come in early to attend that event.

Meetup.com: The concept behind Meetup.com is simply to provide a platform to allow likeminded individuals to get together and talk to each other. I joined the Tucson Meetup vegetarian group, and made an appointment to have dinner at a local vegan restaurant with these folks. Although I wasn't looking to network necessarily, since I was meeting new people the chances of me "networking" were high. And in fact, the guy I ended up sitting next to at dinner was an astronomer at the university and we had a fascinating talk about research. The next time I had dinner with this group

I sat next to a local technology executive who gave me a scoop on an article idea that I could write about for one of my publication clients. So even while we were enjoying the fake fried chicken I still was doing effective and strategic networking that led to a mutually beneficial outcome.

Career fairs: Not only will you meet companies looking to hire, but you'll also interact with other job seekers who have similar goals as you. Whereas someone else might view them as competition and not chat with them, you can take a different approach and recognize that together you can both provide each other networking value. Heck, if the other guy lands the job and you don't, but he knows all about your interests and abilities, he now is your "inside man" at the firm. You might be the next person he hires or engages in some other way, and it all began because you met him at the career fair while you were both looking for work and networking opportunities.

Create Your Own Networking Opportunity!

And of course, there is nothing to stop you from creating your own opportunity to network. This can take place locally or at a conference in a different city. One of the best ways to do this is via social media channels, such as Twitter, LinkedIn groups, and Meetup.com. I have gone to conferences where I have seen people advertise "tweetups" in the evenings, which are simply informal gatherings of people who are at the conference. Have the courage to create your own opportunity to network and you never know what results you will get. And even if only one person shows up, that is a unique opportunity in and of itself, because now you both have each others' full attention and can bond over the humor of having only two people attend. Some of the best events in which I have participated, that delivered the most networking ROI, were attended by so very few people that it automatically fostered a more enriching opportunity to network than if 100 people had showed up. Another idea is to consider launching a professional society or regional chapter based on discipline, career path, or shared hobby, culture, nationality, religious or political affiliation.

Chapter Takeaways

- You know more people and have access to more people than you actually may realize.
- Remember that every professional is looking to solve their problems in new ways, which requires them to network too; in other words, they want to network as much as you do.
- Start your networking plan by conversing with those closest to you, such as family, friends, colleagues, and mentors.

- Networking ROI within your current institution, through professional societies and conferences, is extremely plentiful.
- Don't hesitate to contact people who write articles or who are mentioned in the press – they'll appreciate that you took the time to reach out to them.
- Regional resources, such as economic development, philanthropic, religious, political, and young professional organizations, along with the local daily newspaper and business magazine, are especially useful for understanding the economic and industrial climate in a city and identifying and connecting with leaders and decision-makers.
- Alumni associations are critical and often overlooked networking nodes.
- Spontaneous networking on a plane, train, or taxi can also provide excellent ROI, if you play it safe.
- Be inventive and innovative in thinking of ways to meet new people, network with them, and provide them value.
- This may mean you create your own networking nodes, which is fine – the most successful professionals are usually also the most courageous, and demonstrate boldness in both their craft and their networking that their colleagues take note of and are drawn to.

Notes

i. "Advancing the profession and professional" is the tagline of the American Statistical Association.
ii. http://www.treoaz.org/About-TREO.aspx.
iii. From http://en.wikipedia.org/wiki/Chamber_of_commerce.
iv. From http://en.wikipedia.org/wiki/Business_cluster.
v. From http://tucsonyoungprofessionals.com/about-us/.

7 Networking at an Event

No matter your profession, you'll be attending tons of events throughout your career. And hopefully after reading this book, you'll be inclined to seek out and participate in even more. Below are the strategies to optimize your time and energy before, during, and after an event to ensure you receive the maximum ROI.

Pre-Event Strategies

Recognize that everyone is there for the same reason. No matter what kind of networking event it is, everyone is there with the same objective in mind: You all want to network and meet new people who could potentially progress your career and business in new directions. I make it a point to remember this before I step into any room with a bunch of people. I use it as a confidence booster, because even today after years of hardcore networking, I still occasionally have butterflies in my stomach when I enter the unknown environment of a networking reception. So to counteract this, I concentrate on remembering that everyone is there to meet people and they are probably having a case of the willies too.

Know what kind of an event it is. It's always a good idea to know what kind of an event you are attending. Is this a conference-sponsored mixer, designed to connect early-career professionals with potential mentors? Is it a local "wine and cheese" for leaders in any field? Is it a glow-in-the-dark golf tournament to raise money for charity? Is it a "speed-dating" type of an event with prospective employers? Whatever the affair, discern what type of an event it is and what the goals of the event are in advance. You want to be able to answer these strategic questions:

- What is the flow of the event? For example, is the affair a meet-and-greet centered around an open bar with editors of journals, or is it a small alumni happy hour at one or two tables at a restaurant? This is useful data because you will prepare for and approach each experience differently. If you are attending a sit-down affair you might not have

Networking for Nerds: Find, Access and Land Hidden Game-Changing Career Opportunities Everywhere, First Edition. Alaina G. Levine.
© 2015 John Wiley & Sons, Inc. Published 2015 by John Wiley & Sons, Inc.

as much time and opportunity to meet many attendees because you will be sitting next to the same people all evening. On the other hand, if it is a stand-up mixer in a hotel ballroom, there might be so many professionals in attendance that it might make it challenging to connect with pre-selected people.

- How many people plan to attend? This is useful so you know how the evening will flow, how much time you should spend with individuals, and how many business cards you should bring.
- Is there an agenda for the event, or is it more of an informal gathering? Related to this question, you'll want to pin down what time the event starts and ends, and if there are formal agenda items (like a welcome speech), you want to ensure that you are there early enough to chat with people before the formal part of the affair begins. Often, events have a "networking time" tacked on before or after the scheduled agenda.
- Is there a fee to attend? If you can take care of this in advance, it will help speed up registration when you arrive. And related to this is how much does the event cost, what are the payment methods, and when do you need to pay? If they only take cash on site (which admittedly is becoming less of an issue these days), you'll want to ensure you bring the right amount of money with you. If you have to pay in advance using Paypal, this will alleviate any embarrassment for you were you to show up without paying.
- How should you dress? Obviously, you want to wear garments that are appropriate for the event itself, the organization that is sponsoring it, the venue, the season, the culture of the geographic region, and any activities that will be part of the experience. You also want to feel comfortable and confident in your attire. Many invitations will enunciate the type of clothing they suggest you wear, especially for events for which you have to pay to attend. But if there is no explicit statement regarding suggested attire, contact the organizer and inquire about this. Examples of descriptions of clothing include:
  - Formal attire = dress up a lot: Wear a dark suit, or if you are a woman, wear an evening suit or a fancy dress.
  - Semi-formal attire = dress up a little: Wear a suit, or if you are a woman, wear a suit, party dress, or a nice blouse and a skirt or dress pants.
  - Business casual = neat, well-put-together. According to the Career Services website for Virginia Tech, "Business casual is crisp, neat, and should look appropriate even for a chance meeting with a CEO. It should not look like cocktail or party or picnic attire. Avoid tight or baggy clothing; business casual is classic rather than trendy."[i]
  - Casual = still more formal than what you would wear to hang out at home.

No matter what the suggested attire is, you should always dress a little more polished than you would on a normal day at the office or in the lab. Just as I mentioned in Chapter 4 when discussing professionalism, we want to ensure that others perceive you as a professional and that they don't make snap decisions (especially negative ones) concerning your brand or reputation by your clothing. Your outfit should simply complement your brand; it should not seek to make a statement in and of itself.

> TIP: Your outfit should simply complement your brand; it should not seek to make a statement in and of itself.

- Will food be served? If so, will there be hors d'oeuvres, a buffet, or a plated sit-down meal? Will there be vegetarian options (if this is a concern)? While you are not there for the food, it is good to know whether they will be providing sustenance and in what manner they will be serving it. This will help you determine your course of action in advance and upon your arrival. If there's no food and the event is at 8pm, you'll know to have a meal beforehand.
- Is this an invitation-only affair? Were you one of the selected few who were asked to attend, and if so why?

Research. Research the event (how often do they have it?), the sponsoring organization, and the leadership. Any further information you can learn about the event and its organizers will also help you figure out who else might be attending.

Know what to bring. No matter what type of event, you always want to bring business cards, a small notebook, and a pen, so you can always be at the ready to offer your contact data and jot down notes as you chat with others. It is inappropriate to bring copies of your CV or résumé, unless it is a career fair or another type of event that specifically calls for it. Imagine if you were at a reception and while you were trying to balance your drink in one hand and your plate of food in another, someone handed you a 15-page CV? We don't want to create any awkwardness for you and the other party as we start the relationship, so leave the CVs at home and bring those business cards. However, if the event was advertised as a way to connect employers with prospective applicants, then bringing your résumé or CV (paper copies as well as on your mobile device) is totally appropriate.

When you arrive. Your arrival time will be dictated by the type of event it is. For example, if it is an informal mixer with alumni with a start time of 7:00pm, you can choose to arrive at exactly 7 or you can arrive "fashionably late" like at 7:15 or even 7:30pm. However, being early often has its perks: If you are the first to arrive, you have the ear and attention of the organizer, which can provide you with excellent networking ROI. In addition, as people begin to file in, you will be able to more quickly interact with them since you are there already, you might even ease the nerves of some of the late-comers simply with your presence.

Attending the Event

The following tips apply to mixers and other types of reception where people congregate and interact. However, these same strategies could easily be extrapolated to any other type of an event. Some tips:

- Shut off your cell phone before you enter the event. Your attention should be on the event, not on your phone. If your phone rings while you are speaking with someone, apologize and reject the incoming call and then remember to put it on silent or vibrate. If you have to take calls, excuse yourself and leave the event. You shouldn't be seen on your cell phone at the party.
- Stop in the restroom before you walk in. You want to take one more look at your gorgeous self to ensure that you don't have any spinach in your teeth and that your hair (if you have any) is in place. You also want to give yourself a few moments to relax, gather your confidence, and get ready to have a good time.
- Exude enthusiasm. This is not a chore! This is not a bore! Believe it or not, attending networking events can be one of the best aspects of being a professional in your field. You get the chance to chat with new people who can give you great ideas to solve your problems in fresh ways. You can empathize with them about what motivates and drives you both. You can discuss your passions with like-minded folks who can relate. This should be a fun experience and not an exercise in feeling annoyed or that your work is interrupted. This *is* your work. So as you approach the event and interact with people, put a smile on your face and a twinkle in your eye, but stay authentic. Establish an attitude of excitement, interest, and enthusiasm for being there to share this experience with others.
- Enter the event. Start by getting a nametag (if there is a reception table) and introducing yourself to the person behind the desk. If there is a line to check in, you can start introducing yourself to people in the line. In fact, the line gives you an excuse (not that you need one) to start a conversation.

> **TIP:** When you enter an event, look for "active nexus" – groups of people surrounding something. The congregation, combined with the item that is attracting the people to the area in the first place, is a networking node in and of itself, and allows for effective networking and easy conversation starters.

- Eye the room. There are some people who like to dive into the pool, and then there are others who like to ease into the shallow zone first, get used to the temperature of the water, and then start swimming.

I am of the latter variety. I rarely just dive into a Mixer, but rather start slowly as I acclimate myself to the environment. As I enter an affair, I look around and see if I know anybody. I ascertain where the food and the drink are situated, and where the "action" is taking place, that is, are there large groups of people gathering around something, such as an information desk with literature or a display table? These "action nexus" can serve as a great place to start conversations with people. For example, I attended a charity dinner recently and as I walked into the dining room, I noticed that many people were off to the left. I sashayed over in that direction to find out what all the hullabaloo was about – it turns out that there was a beautiful, large ice sculpture in the shape of a scientific instrument that was not only attracting a lot of attention, but was causing people to congregate nearby to admire it. This "action nexus" had people in it and had a natural subject about which to chat.

- Wander the room and get comfortable. Walk around the room slowly. Look for people you know and for people you want to meet.
- Grab a drink and some food if you like. Although the event's central feature is not the food and beverage, that doesn't mean you can't partake of it. I often like to migrate to the bar after arriving, as it is an "action nexus" in and of itself. In addition, if you are feeling a little nervous, having something to occupy your hands – like a small plate of food or a drink – can also help calm your nerves. I went into detail about food etiquette in Chapter 4, but here are some other tips relating to eating at a networking mixer:
  - Try to keep your right hand free to shake hands, so hold your drink with your left.
  - If at all possible, eschew any foods that are messy, such as spaghetti, or foods that require you to dirty your hands, such as chicken wings or barbecue. You don't want to end up with sauce on your face or your jacket, nor do you want greasy palms. If at all possible, choose foods that only require one bite, such as cheese cubes.
  - Take extra napkins, as you never know what might happen, and you want to be prepared.
  - Once you have your food and drink, migrate to a table or another location where you can put your items down. You don't want both of your hands occupied when you first meet someone.
- Start with people you know. The first people I usually interact with at a party are the folks with whom I already have a relationship. This is partly to ease myself into the event, but you can also view it as strategic: Your friends might be chatting with people you don't know and can introduce you to them. But don't spend too much time with those who you already have a relationship with. Use the time to find new contacts.
- Approach people, especially those who are alone. After eyeing and walking the room, if you don't see anyone you know, but do spy

someone who is standing alone, go up to them and introduce yourself. It is often easier to make conversation with one person rather than a group. And believe me – they will appreciate that you initiated contact! No one likes to be alone at a party and they'll be thrilled that you stopped by.

- Catch someone's eye. As you saunter about, your eyeballs should be fixed not on the floor or the ceiling, not on a pamphlet or on your smart phone, but towards the eyeballs of other people. Your head should be held high, and your goal is to catch someone's eye as a way to launch a conversation.
- Work the room and be seen. Since you are there to launch partnerships with new people, you have to optimize your own time to ensure maximum ROI. So don't spend an hour with one person, or monopolize one professional's time. You can plan to devote 10–15 minutes to each individual you meet. You then want to float around the affair to see who else you may encounter.

Conversation Starters

Starting a conversation is probably the hardest aspect of networking. It is a challenge for most people, both nerds and non-nerds alike – how do you approach and launch a discussion with someone you've never met before? It can be daunting. But don't fret! As you become more adept at networking, you will find that conversations will become easy to launch and will feel like they flow naturally and easily. You won't need to access a brain file of conversation starters because starting the conversation, which is often the hardest thing to do, will become almost innate. But even as you develop this skill, it is still helpful to have at the ready a few ways to commence an engagement with a stranger. Here are a few conversation starters I have successfully employed (and in fact still use today):

- Have you been to this event before?
- Are you a member of (this organization)?
- How are you enjoying the conference/event/day/evening?
- I love your (jacket/bag/hat/bird on your shoulder).
- I see (from your nametag/backpack/shirt/hat) that you are with Texas A&M. What a great school for (volcanology). How do you like it there?
- Nice view, eh? (use this only if there is a view of someTHING as opposed to someone, and make sure it is an actual view as opposed to a wall).
- My favorite: I'm Alaina! As I mentioned above, you really don't need an opening line. You can just start by introducing yourself. Just please use your own name, as opposed to mine.

Another sure-fire way to break the ice with someone is to express gratitude, for something they just did (like a speech) or something that they have recently done (like having written a paper or chaired a committee). I was at a fundraising dinner recently where the two main speakers were both preeminent scientists and as it turned out absolutely terrific public speakers. They each only spoke for about 5–10 minutes, with the goal of exciting the audience about their area of expertise in such a way that they would continue investing in it. Their speeches had a heavy dose of completely appropriate, understandable, and relatable humor; they were goal-oriented and didn't sway from the task at hand; they were energetic and passionate. In sum, the two speakers delivered their messages with such aplomb that the audience gave them each a standing ovation.

As a professional speaker myself, I was extremely impressed with the way they both connected to the audience. After the event, I worked my way through the crowd to each person. I said: "Hi, I just wanted to introduce myself to you. My name is Alaina Levine and I think you just did such an outstanding job speaking. I am a professional speaker on STEM careers and I was really impressed with your delivery. I got some great tips from it – thank you so much for all of the hard work you put into it." The first gentleman smiled and grabbed my hand to shake it and thanked me several times and then started talking to me about how much he enjoyed giving the speech. I gave him my card and asked him if I could follow up with him and he said he would be happy to speak with me further. Our conversation lasted less than five minutes. Then I went to the other scientist and basically said the same thing and he too seemed overjoyed that I took the time to express my appreciation and congratulate him on such a fine job. I got his business card too and plan to stay in touch.

Embed yourself in a conversation. You don't have to wait for an "official start" of a discussion, that is, catching someone's eye or looking for someone standing alone. If you see a group of people that you would like to join, you can stroll over there and casually embed yourself in the conversation. I often employ this tactic, especially at events where I am in truly unfamiliar territory (I am in a new city or a conference I haven't attended before). If I walk into an event and I don't see anyone I know, I will head to the "action nodes" and hover around a spot of people talking. You can do this too and embed yourself.

- Start by acknowledging the people in the group with a smile. But don't interrupt their conversation (yet).
- Use your body language to communicate that you want to be part of the group. Move a little closer to them, lean in to hear the discussion.
- Pay attention to their body language. If they are professionals (as opposed to jerks, see below), or aren't drunk, they will often recognize that you are there to network and will physically shift their bodies so that they have "opened" their circle to you. They may even reach

out and include you in the conversation, by asking you your name or your opinion about a particular subject.

- Listen to their discussion and look for avenues and pauses for you to add value or ask a question. You don't have to wait to be included. If someone says something that is of interest to you, to which you can offer a valuable retort, or creates an opportunity for you to ask a follow up question, take advantage of that moment. For example, I was at a reception in Tucson which took place on the outside veranda of a resort. I arrived at the affair and after doing a survey of the room, I didn't see anyone I knew. I headed to the bar area, which turned out to be a jumbled mess with no clear line. So I just stood in the middle of the crowd for a few minutes and looked to see if I could catch someone in conversation or someone's eye. Before I knew it, I heard the people two inches to my right discussing observatories in Tucson. They were gazing at the mountains in front of us and pondering to each other the distance and driving time it would take to reach the telescopes at the top. Since I was a local and knew the answer, I responded to their query: "It is about one hour or so of drive-time to get to the top of the mountain, where you'll find some amazing telescopes." The couple reacted positively, in that they then asked me some follow up questions and I shared with them information about public programs they could take advantage of since they were out-of-towners. And that was it – I was now embedded in the conversation and the next thing I knew I was having a one-on-one conversation with the woman who had been asking me about the telescopes. As it turned out she was a very successful engineer who had just launched a STEM outreach program at her university. As we talked we both realized we had a lot in common and exchanged business cards so we could follow up and strategize about possibly collaborating.

> **TIP:** Discuss positive things when you first meet someone. Stay on the sunny side of networking and never talk trash about another party.

What to do during the conversation:

- Give them your full attention: Look them in the eyes, smile and nod as appropriate.
- Stop eating, if at all possible – they shouldn't have to observe you chewing.
- Jot down notes: If they are sharing specific information which you know you won't remember, there is absolutely nothing wrong with taking out your notebook and pen and taking notes. And if you don't have a pad handy you can write your notes on the back of the person's business card. The other party may do this as well. But take note:

This practice, while completely appropriate in the United States, is not considered polite in other nations. Know the cultural rules of etiquette before going to a conference or attending an event in another part of the world.

- Ask them questions about their work, research, background, education, current institution. As I mentioned above, get them talking about themselves and get them talking about ideas that bring them joy.
- Listen most of the time and respond to their answers in a way that demonstrates you have been paying attention.

Conversations you should stay away from:

- politics
- religion
- divisive issues
- inappropriate and/or offensive language (i.e., don't drop an F-bomb the first time you meet someone, even if they do)
- inappropriate and/or offensive humor.

And probably most importantly, you should never engage in a conversation that refers to another party in a negative way. You should never be seen as talking trash about someone behind their back.

When to exchange the business card: You can ask for and offer your business card at various times in the conversation, such as:

- At the beginning: "Hi my name is Alaina Levine. I am a science writer." (Hand them your card.)
- In the middle: "That's a very good point about postdocs and career paths. Let me give you my business card. I am a career consultant with an expertise in helping postdocs define and land their dream career."
- At the end: "It was such a pleasure speaking with you. May I have your business card? I would love to follow up with you in the near future. Here's my card."

Conversations Enders

This is one aspect of networking at events that perplexes even the most seasoned professionals. Once you engage someone in a conversation, how do you end it? Knowing "when to say when" is tough to gauge. But here's one You don't have to wait until the conversation comes to a formal end, or when you both run out of things to talk about. You should certainly wait for a pause – don't interrupt them and yell "bye," or slip off when they look away like that astronomer I encountered in Chapter 2. Here are some effective and polite ways to end a conversation, when there is a pause:

- I really enjoyed our conversation. Do you mind if I follow up with you next week?
- It was a pleasure meeting you. Can I get your business card?
- Thank you for the opportunity to chat. I would love to continue our conversation.
- It was great speaking with you. Enjoy the rest of the event/conference/parade/charity gala ball.

And then extend your hand, shake theirs, and be gone! You have more networking to do (as do they), so get on to the next "action nexus" or opportunity that allows you to make someone's acquaintance.

If the Person is a Jerk

A shocking and little known truth about STEM fields is that sometimes, very rarely mind you, they attract people who are jerks. You know these folks – they are rude, they talk about people behind their backs, they are insulting to you and others, they tell racist or sexist jokes, they want to hug you when they first meet you and hold on for a super long time. They are all-star idiots and the depth of their idiocy may know no bounds. It is important to make a distinction between a jerk and someone who is not as adept at networking as you:

- A supreme jerk is someone with whom you would never be able to craft a valuable collaboration. Their attitude and their brand is one of negativity and they treat people rudely and with complete disrespect. Note that even some of the most celebrated scientists and engineers are jerks.
- An inexperienced networker, who doesn't know the standard practices and etiquette when interacting with someone at an event. They mean well and they don't want to cause you harm but they don't know the social norms of networking. They probably are not overtly insulting, and their behavior demonstrates their networking naiveté.

Warning: Sometimes, you will not be able to tell the difference between a jerk and an inexperienced networker. Their behaviors can certainly overlap. The astronomer who ran away from me after the Big Bang occurred at our networking event? He was not a jerk. He was someone who was very unfamiliar with social decorum and probably felt so uncomfortable in that situation that he was thrilled he had an excuse to high tail it out of there.

But here's an example of someone who was a true jerk: I once attended a nice party in recognition of a committee on which I volunteered. I was mingling and migrating through the affair looking for someone to speak with when I saw Dr. X and Mr. Y chatting. I had admired Mr. Y for many years; in fact, at one point I had applied to work in his division. We had never had the chance to meet personally so I was excited to be able to catch him at the event.

I approached Dr. X and Mr. Y who were standing in such a way that they were facing each other directly. I hovered nearby, smiled at them both and leaned in to the conversation, expecting them to turn towards me and with their body language invite me to participate in their conversation. But they did the opposite. They actually turned away from me. I was new to networking and didn't know what to think of this, so I simply inched myself a little closer to the duo. They proceeded to turn their backs completely to me while continuing their extremely important and critically private conversation.

That action was no accident. They knew what they were doing. They were behaving like jerks. And after I digested what I had just witnessed, I realized how grateful I was that I was not collaborating with either Dr. X or Mr. Y.

Even though you are not a jerk, this is an important reminder that people are watching your actions, judging you based on your behavior and attitude, and making on-the-spot decisions about whether they should engage you further. Don't be a jerk. Don't even give a whiff of being a jerk. Be the classy professional that you truly are.

But if you encounter a jerk at a networking mixer, there is no reason to continue hanging around with him or her and wasting precious time that could be spent engaging someone who is a prospective partner. Wait for a slight pause in them speaking, and then you can simply state:

- Well this has been great speaking with you. I am going to get a drink. Enjoy the rest of the conference.
- Would you excuse me? I just realized I have to make a phone call.
- So nice to meet you.

If they are more than a jerk

Certainly if the person is being rude and offensive you don't have to stand there and take it. You can simply extend your hand if you care to, bid them farewell, and walk away. If you feel you want to make a stand and call them out for their offense, especially if they are saying things which are sexist or racist, then don't be afraid to do so. A simple "I find that offensive and disgusting" will suffice, and then just walk away. My point is this: If they are demonstrating that they are a more than a jerk, then you don't have to maintain the same level of decorum that you would normally maintain with others. I once was at a mixer where someone (who more than likely had had a few too many drinks), whom I had just met, decided that it was just fine if he shared a particularly offensive sexist joke. Not only did I not find it funny, but I was shocked by his use of filthy language in that situation. My response? "Ok then," I said, and walked away. Needless to say I never spoke to him again.

Following Up

Finally, as I have discussed throughout this guide, it is not enough to say hello, chat, bid them farewell, and never ever speak to them again. You must follow up to keep the prospect of a partnership alive and to ensure that you (and the other party) can gain access to the Hidden Platter of Opportunities.

Chapter Takeaways

- When attending an event, recognize that everyone is there for the same reason: To network.
- Do your research ahead of time to find out the specifics of the event so you can be prepared and at your most confident.
- Practice your brand statement and be ready to discuss it and ask questions of other parties.
- Look for an "action nexus," a kind of networking node where a congregation of people are surrounding an area or object. This can serve as a conversation starter and ease you into the networking scenario.
- Optimize your time – you and everyone else attending only has a short window in which to network, so don't spend 45 minutes with the same person (unless it is Bill Gates!).
- Don't be afraid to go up and just introduce yourself to others – you really don't need an opening line.
- Exude confidence and show off your authentic enthusiasm and excitement for attending the event.
- Keep the conversation positive and stay away from topics that are divisive or potentially offensive or upsetting.
- Watch out for jerks, networking novices (who don't necessarily know what they are doing), and people who are even worse than jerks.
- Follow up and stay connected.

Note

i.   From http://www.career.vt.edu/JobSearchGuide/BusinessCasual Attire.html.

8 Social Media Networking

Networking via social media channels is no longer an option; it is now a required and strategic element of any successful networking campaign. But just like "real-world" networking, there are various tactics, guidelines, and tips to utilize and employ to ensure that your social media activities deliver excellent networking ROI.[i]

We know networking is a necessity for career advancement in science and engineering. And social media networking is no different. Your online presence, via websites, your blog, and personal profiles on channels such as Facebook, LinkedIn, and Twitter all serve as a way for interested parties to get to know you and your brand, and to learn how you might contribute to their enterprises. And it is becoming more and more critical for you to maintain a presence on social media in order to build digital trails that amplify your reputation and reach decision-makers in your field.

Just like all networking strategies, we must start with our goals and determine why and for what we would use social media tools for networking:

- Gain access to Hidden Opportunities.
- Gain access to new networks.
- Learn who are the trend-makers and decision-makers.
- Connect with decision-makers
- Gain insight about a community, field, job market, hiring trends.
- Find a job, grant, fellowship, other opportunities.
- Gain inspiration and new ideas.
- Amplify your brand, attitude, reputation.
- Promote yourself.

When you apply for a job or fellowship, or send a cold email to someone, one of the first actions that the other party takes is to Google you. And the second action they take – which is increasingly becoming the norm – is to check your LinkedIn profile. If you don't have one, the perception might be that you are not a contributing member of your community. In fact,

Networking for Nerds: Find, Access and Land Hidden Game-Changing Career Opportunities Everywhere, First Edition. Alaina G. Levine.
© 2015 John Wiley & Sons, Inc. Published 2015 by John Wiley & Sons, Inc.

> **TIP:** If you don't have a LinkedIn profile, the perception might be that you are not a contributing member of your community.

some recruiters have told me that they envision that the LinkedIn profile will soon supplant the résumé as the standard for finding qualified applicants for job openings. In some industries, it probably already has.

There are currently 300 million users[i] on LinkedIn, and don't think that it is only for those who work in non-academic jobs. Increasingly, more and more academics are on LinkedIn, including professors, postdocs, and even students. And academics are using it for the same reasons non-academics are: To increase the size of their networks and develop inroads into new ones. As such, I recommend setting up a LinkedIn profile extremely early in your career; you should definitely have one in grad school, but I don't think that having one as an undergraduate is too early or inappropriate.

In fact, let me be extremely blunt: *you absolutely must have a LinkedIn profile* to advance in your career. In the last six months alone, at least 10 people whom I know personally found employment *specifically and only* because they were on LinkedIn. Most of them had their profiles found via decision-makers who were conducting searches for professionals with certain skills and expertise. My other pals either discovered jobs that were only listed on LinkedIn and applied for them, or used their LinkedIn Connections to develop relationships with insiders at the organizations for which they wished to work. In all cases, my friends' LinkedIn profiles paved the way to crafting relationships that resulted in landing employment. Social media networking doesn't get any better than that!

Developing a social media networking strategy involves planning and preparation, execution, and monitoring and maintenance. You can't simply post your profile on LinkedIn or send one tweet and then sit back and passively expect to reap the rewards. You have to consistently engage and interact with others, and you have to develop a plan that optimizes your time while seeking to achieve your career goals.

You also have to understand which social media channels are the right ones to utilize in the professional ecosystems in which you dwell. Unlike

> **TIP:** LinkedIn is a professional marketplace to exchange information of value.

Facebook, which can be used for both fun (see the "Jean Luc Facepalm Facebook page") as well as professional activities (many companies including my own have a Facebook page to converse with fans), LinkedIn is meant to be a purely professional communication channel. This means that while on Facebook with your friends you may share pictures of felines in compromising poses, on LinkedIn you only share information, ideas, and connections which are related to your industry and demonstrate your professionalism. Think of LinkedIn as a professional marketplace to exchange information of value. And again, while cats are adorable, there is no professional value in sharing a picture of a cat with a scientist.

Principal Pillars of Social Media Networking

Your activity on LinkedIn will serve as the cornerstone of articulating your value to the public, and in fact many of the principles associated with leveraging LinkedIn for networking and career success are universal and can be applied to other sites. You will probably utilize other platforms in which to communicate with your publics, like Facebook and Twitter, as I discuss below. There are even more specialized channels such as Research-Gate or Frontiers, both of which are becoming more popular in the STEM community. But before you post anything anywhere, consider these Seven Principal Pillars of Social Media Networking:

1. Be professional, at all times, in every way and on all channels: Social networking has a specific purpose and part of that is to increase your connectedness to other professionals. It is vitally critical that you are perceived as a professional, so make sure all of your activities – your postings, comments, questions, pictures, and videos – all reflect your professionalism and dedication to science and engineering. This means using appropriate language and not posting things that reflect poorly on your reputation and brand, such as offensive or divisive comments, or inappropriate or silly pictures of you and cats.
2. Be dynamic: Seek to engage the members of your community with content that is relevant to them, by being both *proactive* and *reactive* with your activities. Be *proactive* by posting new content to your social media groups, starting new discussions, and contacting people within the community whom you find interesting. Be *reactive* by commenting on what others post or contacting people who post things of interest to you to continue the conversation via an informational interview.
3. Be valuable: Everything you post, whether it is reactive or proactive, should be construed as a value addition to the community. By adding value and not wasting people's time with information you accomplish two tasks: You elevate your brand and reputation in the minds of those around you and you are perceived as a credible leader and expert. And therefore more people will be interested to hear what you have to say. Furthermore, once you are established as a thought leader, others will seek you out to establish potential partnerships. It is all about delivering ROI for those who take the time to read your contributions: Make sure that every time someone sees that you have posted something, they know it will be incredibly worth their while to read what you have offered.

4. Be seen: The more you engage others with professional, dynamic, and valuable content, the more you will be seen as being professional, dynamic, and valuable. Every action on social media contributes to the public's perception of your brand, attitude, and reputation, so post often. Build a buzz around your reputation so much so that those around you can't *help* but think of *you* for that next great (often hidden) career opportunity.

5. Be consistent: Since your brand is your promise to deliver excellence, dependability, and expertise in whatever you do, it is important to ensure that all of your social media interactions are consistent with your brand. This also involves making certain that your brand is consistent across platforms. For example, your user names on different sites should either be the same or very similar, so that when people see a post from you, they know it is YOU and not your doppelganger.

6. Be aware and mindful that anything you post is accessible by anyone else from now until the end of time: Although this seems to be a notion steeped in common sense, it certainly merits specifying and amplifying that once you post something it is online forever. As my mom says, "cyberspace is foreverspace." And if you think that because you tweaked your "privacy" settings only a handful of people can view your rant or rave, remember that on the internet there really is no such thing as "privacy." So before you comment on or post something that is emotionally derived, fraught or driven, ask yourself "how will this statement contribute to my community and to my advancement?" and consider that possible partners, decision-makers, current and future colleagues and supervisors, and those with the likelihood of offering you access to the Hidden Platter of Opportunities, will see it. And when they do, if it is at all undesirable or offensive, any chance you have of collaborating will disappear.

7. Be careful that it doesn't become a time sink: You probably have heard about people who admit that Twitter, along with other social media sites, can very quickly become a time sink for

> **TIP:** Don't let social media become a time sink. Use it as a tool for networking and be careful that it is not at the expense of you producing excellent work.

someone if they are not extremely careful. If you are on social media, you probably have had the experience of constantly checking the feeds that emanate from your accounts on LinkedIn, Facebook, and Twitter, and then before you know it, three days have gone by and you have been absolutely non-productive. Social media is a tool for networking, but it cannot take priority over your professional outputs. So use social media carefully and pay attention to how much time you spend reading posts, responding to comments, tweeting and reading others' tweets, and so on. Remember, just like any form of networking, if you invest so much time in the activity, that it is at the expense of you producing excellent work, then it will be for nothing. Your job and profession must come first.

Preparation

To launch your social media networking plan, begin by researching what is online about you. Google yourself and various versions of your name, on a regular basis. For example, I regularly Google "Alaina G. Levine" and "Alaina Levine." Go through numerous pages – don't just rely on the first five results, as there may be information about you online that only comes up in Page 10 of your results. And you might be very surprised what you initially find.

As I mentioned above, when I was an undergraduate, I was heavily involved with the physics department and participated in the Society of Physics Students (SPS). During my junior year, the department began the process of selecting a new head and SPS was granted the rare opportunity to interview the candidates and contribute to the final recommendation to the Dean. I thought it was pretty significant and unusual that faculty would support such an endeavor, and apparently so did the school newspaper, because the editor assigned a reporter to write an article about the students' role in the interviewing process. When the writer asked me what I thought about it, I said something to the effect of "Right now, the department is at a crossroads. It can get better, it can go to hell, or it can stay the same, and the students are helping to determine the course of action." So what do you think the reporter chose to quote me as saying? "'The Physics Department can go to hell,' said mathematics junior Alaina Levine." To this day, if you conduct a Google search, and weed through enough pages, you can still find that quote.

It is better for you to know now what is out there about you, including the good, the bad, and the exceptionally ugly, rather than a potential employer or partner finding it themselves. This is especially critical if you find that there is misinformation about you personally, or there is someone with a similar name who is doing things that could potentially damage your reputation.

> TIP: Google yourself regularly to keep track of what is online concerning you.

Make it a habit to not only routinely Google yourself, but also to add search terms which prospective collaborators may use to find you. For example, if you were a microbiologist with an expertise in undersea microbes and ecology, you should use these as search terms and see what you come up with. We want to ensure that your online presence not only allows you to communicate data about yourself, but that it leads people back to you – they should easily be able to find you if they need someone with X and Y skills who has a background in undersea microbial ecology.

> **TIP:** If you find something online about you that could potentially damage your reputation, whether it is true or not, you will need to take immediate action to cleanse your brand and save your reputation.

Furthermore, don't limit your self-searches to Google. Use the other search engines such as Yahoo! Search, Bing, and Dogpile to see what is online about you. You can also go to specific sites, such as LinkedIn, Twitter, or Facebook, and do searches within those arenas.

In addition to researching your own digital trail, you also want to start exploring what social media platforms would be most useful to you and deliver the most ROI in terms of networking and accessing hidden opportunities in your career and discipline. LinkedIn should be at the top of that list, but Twitter and Facebook are also relevant to the scientific and engineering communities (as I specify below). Moreover, as Google+, Pinterest,

> **TIP:** Ask your mentors and colleagues with which social media channels they engage.

and Instagram gain additional users and activity, these may also be channels on which you would like to have a presence. Finally, there are research-related online networks as I noted above, such as ResearchGate and Frontiers, sites organized by professional societies devoted to specific STEM areas, and even a site organized by my publisher, Wiley, that may be of interest to you.

And while we are discussing your online presence, keep in mind that you want to craft as many positive digital trails as possible that allow decision-makers to find you. So in addition to having a profile on social media sites, don't discount the value of posting your résumé on certain strategic websites and job boards that are read by leaders in your field. Some professionals may get so caught up in creating the perfect LinkedIn profile that they forget the simple act of posting their CV on the career center website of their professional society, for example. But this can be an extremely important way for you to access posted positions, and more significantly, for others to find and offer you access to the Hidden Platter of Opportunities.

> **TIP:** Craft as many positive digital pathways as possible that enunciate your brand and that allow decision-makers to find you to offer you the Hidden Platter of Opportunities.

A colleague of mine recently started a new job in risk management with an international company. She absolutely loves the position. She did everything right to ensure her brand was visible in the virtual world – she had an excellent LinkedIn profile that clearly communicated all of her skills, she engaged other professionals on LinkedIn and other sites and sent cold emails and arranged informational interviews with potential leads. She also posted her résumé on the online job board for her professional association. And one day she received an email out of the blue from a recruiter with the firm stating that he had found her résumé, was impressed by her unique blend of skills, and queried whether she would be interested in a job with his organization. The position may have been advertised, but the hiring manager personally reached out to her because she was present where he was looking for talent. As such, he offered her the hidden opportunity to apply for the position ahead of anyone else who may have independently found the job on the company's website. This immediately gave her the competitive advantage.

Building your LinkedIn Presence

TIP: Listing your awards and honors on your LinkedIn profile (as well as your résumé/CV and other marketing documents associated with career planning and job searching) is NOT bragging. You are authentically articulating important information to other parties that enables them to understand the significance of your achievements and extent of your expertise in order to make a decision to engage you.

LinkedIn will serve as the foundation for your other social media actions (and for now, it will probably be the site you spend the most time on because the potential networking ROI is the highest). Therefore, it is important to have an eye-catching, easily-searchable, easily found profile that plainly articulates your value, skills, and expertise. Just like most of the other social networks, you can join LinkedIn for free. And making a profile is very easy to do. The profile serves as a "living résumé" and describes your current goals and professional interests, current and previous jobs, technical and business abilities, education, projects, publications, and other relevant information.

Your profile should include:

- Your name: Use the same name as you do on all of your social media sites, CV, institutional website, journal papers, and so on. In other words, if you are known as Carla Annette Martinez, seek to keep this consistent across the channels (as one of the Pillars above reinforces).
- A title that communicates your brand and expertise: This should be a positive title, something that catches people's attention. You don't have to list "Unemployed Mechanical Engineer," but rather you could

write "Mechanical Engineer, Technical Writer, and Researcher in X." My LinkedIn title is "President, Quantum Success Solutions: Speaker, Consultant, Writer, Comedian, Entrepreneur; Author, Networking for Nerds (Wiley, 2015)." I have seen profiles that even give the number of years' experience someone has in a particular field, or list subfields that may be of interest to potential partners. In a certain respect, the title is an abbreviated version of your brand statement, as I discussed in Chapter 2, so make it pithy, meaningful, and exciting so that if someone reads it, they want to read more about you.

- A professionally-taken picture: This should be a headshot and allow the viewer to see your eyes. According to Yumi Wilson, Community Manager for LinkedIn for Journalists and Friends of LinkedIn, a profile photograph increases your visibility by seven times![ii]
- Your industry: You can list your industry as a part of your profile and LinkedIn allows you to choose from a list of popular industries, including Research and Education Management. This is one more way that someone can search for and find you on LinkedIn and elsewhere, so think carefully with which industry you want to identify yourself.
- A customized URL: It only takes a few minutes to convert your ordinary LinkedIn profile URL, which includes a seemingly random jumble of numbers and letters, into a customized one that includes your name. This is a strategic action, because you want to be able to easily distribute the URL and drive traffic to your LinkedIn profile. With a customized URL you can easily point people to it and include it on your CV, in your email signature, and even on your business card.
- A personal email address: When you sign up for LinkedIn, consider using your personal email address (i.e. not the one associated with your institution). The reason behind this tactic is that if you create an account using your email from your current organization, when you graduate or leave, you might lose access to that email and thus lose access to your LinkedIn profile. By having a personal email address as either your primary or alternative email in your LinkedIn account, you can avoid this scenario.[iii]
- Summary: Second only to your title, your summary is the most important element of your profile, because it is the first opportunity you have to go into some detail about your brand. So just as your title is an abbreviated brand statement, your summary is a more detailed discussion of your value, skills, and expertise. Everyone has a different way of writing their summary. Some people list bullet points, others use paragraphs, and still others use a combination. Some people list the number of years' experience they have in a particular field, as well as their desire to work in a certain sector or discipline. The most efficient way to craft your summary is to start with your brand statement, and think about what you want to market about yourself to potential partners. What is the strategic information

that you can provide that would convince someone they would want to offer you access to hidden opportunities? What value can you offer allies? What are your specific problem-solving abilities? Take a look at the profiles of some of your mentors and leaders in your field to get some good ideas as to how you can organize yours effectively and what type of data would be relevant and clever to include. Want some examples of clever data? For people focused in on academic jobs, I would list how many grants you have received (and the total amount funded), papers published, and talks given, and any other metrics that higher education decision-makers consider important for employment decisions. Similarly for non-academic positions, I would include the number of teams I had overseen, total budgets I was responsible for, total sales I achieved, and so on. And of course, if you are interested in careers in both arenas, you can certainly include all of the above, as it is all relevant.

- Experience: The section will mirror your résumé. List any relevant jobs or leadership positions you have had, including volunteer experiences, in reverse chronological order. And then communicate your accomplishments for each position by clarifying in each bullet point the following:
  - The problem you were given.
  - The solution you came up with.
  - The result(s) of your solution, with quantification if at all possible. For example, "Designed a strategic communications and outreach campaign designed to promote the department to teachers and K-12 students. Results included a 100% increase in attendance at departmental events and a $5M grant to expand the reach of existing programs." The reason behind this is simple: Since the purpose of every job is to solve problems, it is imperative that you communicate what problems you have solved, how you solved them, and what results you gained. And by quantifying the outcomes, you provide context and significance to your achievements. This clearly enunciates your value to the community, and allows others to start contemplating how you can solve their problems and make their life easier – a sure way to be offered both promoted and clandestine opportunities.

  Take note that you probably won't be able to list every achievement via problem–solution–result, but do your best to present your information in this format, as it makes the readers' decisions easier as to whether to engage you.

- Skills and expertise: LinkedIn makes it very easy to qualify your skills, because it offers you a list of skills from which to choose. The advantage of this method is that since the wording of the skills is uniform, recruiters and others who are looking for partners with that specialty can do extremely efficient searches to find the professionals to fill their needs. For example, if one of your skills is oral communications, LinkedIn offers you the opportunity to list it as "public speaking."

So if I was looking for a chemical engineer who had experience in public speaking, I could easily find you. You don't have to worry that someone might type in a slightly differently-worded search criterion and then not discover your profile. When you edit this section, you can search for relevant skills, and you can list up to 50 of them. And you can edit your list of skills depending upon what type of opportunity you are seeking during different phases of your career.

- Endorsements: Once you have your skills listed (and after you start adding Connections (see below)), LinkedIn's algorithm will start asking your Connections to Endorse you for the skills you have listed. As people endorse you, their pictures get listed next to your skills, along with the number of people who have endorsed you for that skill. The most endorsed skill in your profile rises to the top of the list and so on. The endorsement process is a clever visual marker for how others perceive your expertise. Although there hasn't been any comprehensive study as to whether the number of endorsements aids professionals in landing jobs, it does demonstrate to viewers of your profile that many other people consider you an expert in certain areas. It is just one more way to visually boost your brand in cyberspace.

- Honors and awards: Don't be shy. Trumpet your awards. Let people know you won X Honor or Y Award. You can even add a line or two about its significance, which is a good idea; if someone is not familiar with the award they may not realize that it is a Big Deal. So if you won a fellowship that is only given to two people in your entire country each year, say so.

- Education: Depending on where you are in your career, your education may be placed on your profile higher or lower. Regardless, you definitely want to include your alma mater(s), degree names, degree subjects, and what year you received them. You can additionally list coursework or add a few lines about your dissertation/thesis, such as its title or the subdisciplines you researched.

- Other sections include:
  - projects
  - papers
  - languages (this is very important as you look for partners across the globe and to distinguish yourself from others who only speak one language)
  - patents
  - certifications
  - organizations
  - volunteering and causes.

LinkedIn allows you a lot of flexibility – you can change the order of items in your profile, you can pick and choose what parts of your profile you want visible or hidden from the public, and you can even add videos, pdfs, and powerpoint slides for enhancement.

> **TIP:** On LinkedIn, only connect with people with whom you have had a "meaningful engagement," such as a conversation, an email exchange, or an informational or job interview. Generally, don't invite strangers to connect, and conversely, don't accept invitations to connect from people you don't know.

Establishing your LinkedIn Profile is only the first step in optimizing your brand communications via this social media channel. There is still much more to do and much to be gained. Once you put together your profile, the next step is to start adding connections.

"Connections" on LinkedIn are similar to "Friends" on Facebook – they serve as a list of your contacts, and you can almost use LinkedIn as a rolodex to keep track of your professional acquaintances. Since LinkedIn is based on six degrees of separation, you can leverage this platform to reach even more people and organizations in which you are interested.

There is an art to inviting people to "connect" on LinkedIn – in general, don't invite strangers or people with whom you haven't interacted at all. Part of the reason for this is because once they are connected to you, then they are connected to all of your connections too. You essentially are vouching

> **TIP:** If you go on a job interview that seemingly goes well and you don't get the job, connect with the decision-maker on LinkedIn. After all, if you are still on good terms with them, they are now a part of your network, and vice versa.

for them, and you don't want to put yourself in a position where you are vouching for someone who you don't know. Similarly, you wouldn't want to give a stranger access to all of your colleagues' information. So be careful with accepting invitations to connect. I tend to invite people to connect or accept invitations to connect with people with whom I have had a "meaningful engagement" – at the very least a conversation or an email exchange. And I very rarely accept invitations from others to connect whom not only do I not know, but don't even bother to personalize their invitation message to me (see below for an exception to this rule).

Some thoughts about adding connections:

- As a student, should I add my professor? If you are an undergraduate or graduate student and you are taking a class, wait until the course is over before you invite the professor to connect.
- Should I add my Principal Investigator/Boss? Although more and more people are adding their supervisors as connections, this action is up to you depending on your comfort level and the nature of the

relationship with your boss. Just remember that once you add them as a connection, they will have access to your other connections. On the other hand, since LinkedIn is a professional marketplace, it is generally appropriate for one to invite their boss to connect.

- If I go on a job interview, should I add the hiring manager? You certainly can do so, but I would wait until the interviewing process has been completed and a decision has been communicated to you. You don't want to give the other party the opinion that your major concern is adding connections when it should be a complete focus on the job and organization. But after they offer you the position, you can certainly invite them to connect. And if you don't get the job, fear not – you can still invite them to connect and in fact, you absolutely should do this. They are now part of your network.

- If there is someone with whom I really want to connect, but have no ties to them, how can I do so professionally and appropriately? If you are looking for people on LinkedIn and happen upon the profile of Dr. X, someone who you don't know personally, but who you admire and with whom you think there may be an opportunity to collaborate, you can send her an Invitation to Connect with a personalized message such as "Hi Dr. X, I really enjoyed your paper concerning dark energy in Science Magazine. I would love to speak with you to discuss a potential collaboration. Thanks, Alaina." This message articulates that your aim in connecting is not to just add as many connections as possible, but rather that you want to "network" – to build a win-win partnership with this professional. The default "Invitation to Connect," which states "Hi Alaina G., I'd like to connect with you on LinkedIn," or "Please add me to your LinkedIn network" tells me nothing about yourself or more importantly, why I am someone with whom you desire to align yourself. Just as with cold emails, you have to customize the Invitation to Connect if you are to get any traction and craft lasting relationships. Please note that using personalized Invitations to Connect are my only exception to the rule mentioned above concerning not to add people to your connections that you don't know. Even so, this method has its problems – unlike a message that you send through LinkedIn, an Invitation to Connect does not allow the other party to easily send a response to you. So while you may be able to connect with them, you would then have to send a follow up via a private message on LinkedIn or email them asking for an appointment to speak. A far more functional avenue to adding connections is actually to send someone a cold email first, chat with them, and then Invite them to Connect. I have found this method elicits far more responses and also solidifies far more relationships than making the first interaction an Invitation to Connect, even if it is a personalized one. However, if you don't have the party's email address, then certainly either personalize the Invitation to Connect, or send them a private message if they have posted something in a Group (see next page).

One of the best features of LinkedIn connections is that you can search for people via their company or institution, alma mater, location, industry, or discipline, and LinkedIn will tell you how many degrees away you are from that person and who the people are in between. This allows you to ask for introductions from your connections. I have found this tool to be extremely useful in accessing new networks and even landing gigs.

LinkedIn Groups

Other than your connections, your activity within LinkedIn groups is the most important method you will use to network effectively and engage collaborators. Join groups that relate to your interests and career goals, and which can provide you access to new spheres of influence. At the very least join the groups associated with your alma mater(s), current insti- tution, professional society, a conference that you regularly attend, and regionally-focused associations tied to business enterprises (like those I discussed in Chapter 6). And there are other groups that relate to subfields, career tracks, and particular segments of society, such as the National Postdoctoral Association, Wiley Job Network, and Astronomers Beyond Academia. I host a group called "Alaina's Alumni: Career information for scientists, engineers and non-nerds too." You can join up to 50 groups.
Some tips related to groups on LinkedIn:

- *Groups are the most important feature in LinkedIn*: They provide you with instant access to like-minded souls with whom you can build alliances and who can offer you access to the Hidden Platter of Opportunities. Take full advantage of groups and use them as your main avenue for networking.
- If you are unsure of what groups to join, take a look at the profiles of professionals you admire and see which groups they belong to.
- You can choose whether you want your groups visible to others who view your profile. This may be an important option for you, especially if you are looking for a job while still employed. You might not want your boss to see that you belong to groups that are associated with jobs, such as the group Science Jobs. So you can edit your profile to hide cer- tain groups or all of them if you wish. But keep in mind that even if you edit your profile so certain groups are not listed, if you post something in those groups, the posting is not concealed. Generally, someone can always find that you posted something to the group.
- Review who is in the Group: Once you join a group, you have access to the full directory of everyone who is also a member. You can use this directory as a way to connect with people.

- Take advantage of jobs that are promoted in the groups. When a hiring manager advertises a job in a group they do so very strategically. They know that this group attracts people with the talents and experience they desire. So certainly if a job that is listed appeals to you, apply for it. But even if the job is not attractive to you for whatever reason, you can still contact the hiring manager and ask for an informational interview. They have made themselves known in this forum and you can certainly take advantage of that to forge a potential networking partnership that could lead to a job (or other career opportunity) that is more appropriate for you in the future.
- Follow the Seven Principal Pillars of Social Media Networking, especially #2 (be dynamic):
  - Be proactive: Post new stories, links to articles and videos, questions, and ideas for the community. You will be seen as a thought-leader and a trend-maker. The more you start conversations, the more visible you will be in the group and the more people will think of you for hidden opportunities. I can't even begin to count the number of opportunities I have been granted (both hidden and promoted) simply because I continuously posted valuable content to a specific group and thus people perceived me as someone with whom they wanted to do business. And then they reach out to me to launch an exploration of a win-win alliance.
  - Be reactive: Respond to queries, add new information to discussions, and be seen as a problem-solver. And you don't have to post just once and be done. You can follow the conversation in the group (via settings that email you when someone adds to the thread) and if you have new insight, share away.
- Pay attention to opportunities to reply privately: When you follow a thread in a group, you have the option to comment publicly or privately. If you choose the former, your response is displayed as part of the thread itself and is visible to everyone in the group (and potentially anyone who does a search for your or the group's activity). There are advantages to this, as discussed above. However, if you choose the latter, you can send a private message through LinkedIn to the poster that only they will see. This private communiqué can deliver terrific networking ROI. I often take note of discussions in groups and when I see someone with whom I am interested in connecting has either started the thread or is commenting on the thread, I send them a private message, complimenting them on their comments and asking for an informal conversation to discuss the subject further.
- Arrange your LinkedIn settings so you receive a daily digest of activity in groups. You'll find out about up-to-date information relevant to your career and networking interests and you'll also be able to track your own activity.

Other features of LinkedIn that will assist you in your networking efforts:

- Seek recommendations: Just as endorsements enhance your profile, so too do recommendations. In fact, recommendations are actually more important than endorsements because someone actually had to take the time to craft a statement that specifically states why you are a valuable asset to their team. Ask your mentors and those who have observed your success to write short recommendations for your profile. Go for quality over quantity and if they so desire, let them know you'd be happy to send them a draft to help them get started. And the beauty of recommendations is that they are verifiable – they cannot be faked. Someone in LinkedIn, specifically one of your connections, had to write, authenticate and allow you to post the recommendation on your profile. Check out "Who's Viewed Your Profile": What a great aspect! You can actually see who has been scoping you out, and if it is someone with whom you want to connect, now you have the opportunity to do so and an excuse to make contact. On many occasions, I have reached out to professionals who I noticed had checked out my profile, with the simple statement of: "Hi Sydney, I saw that you had looked at my profile. There definitely seems to be some synergy between our areas of expertise and our projects. I would love to chat with you to learn more about your work and see if there is an opportunity to partner. When in the next few weeks would you be available for a short conversation?" More often than not, I get positive responses and a conversation commences. Take note, however: Unless you upgrade to a premium account (which requires payment versus the free account that anyone can get), you will only see a small sample of the number of people who looked at your profile. You have to upgrade to get the full list. But I recently heard from a postdoc who did this and paid the monthly fee for a higher tiered account and was able to secure job interviews just from taking advantage of the "Who's Seen Your Profile" feature. It really is that useful. See below for more information about the fee-related accounts.
- Introductions: One of the best ways to pay it forward to your net-works is to offer to (or when asked) provide introductions to others in your network. LinkedIn provides a special way of doing this. For example, let's say you were interested in connecting with someone from McKinsey & Company. You could start by doing a basic search on LinkedIn for McKinsey and you would get a list of people who are in your network or are two or three degrees away from you who have McKinsey in their profile. You can even limit the search so that you obtain a list of only current McKinsey employees as opposed to those who have worked there in the past, and/or McKinsey employees who were educated at your alma mater. In your search results, you notice

that Jane Smith works for McKinsey and LinkedIn tells you that she is two degrees away from you and even tells you who are the people in your network that are connected to her directly. Then you can either "connect" with her directly or "ask to be introduced" by one of your in-common connections. The former won't necessarily provide you with a response (as discussed above), but the latter often yields positive results. And furthermore it gives you a chance to reconnect with the person who made the introduction. And by the way, if the introduction does yield a result and you get in contact with Jane, make sure you send a thank you email to the person who introduced you.

- Updates: You can add updates about your activity which then get broadcast to your connections and are also visible on your public profile. This is an effective way to communicate your latest achievements, or to send out queries for assistance or ideas. The key thing to remember with this feature is to use it only for professionally-related communications that deliver some value to others. In other words, it is not to be used to articulate your love of a certain ham sandwich.

- Search for Alumni: If you hover over the button marked "Network" on the main menu bar at the top of your LinkedIn profile, you will get an option of "Find Alumni." When I first learned about this, in the course of writing this book in fact, I couldn't believe how powerful this feature is. You can search for alumni of any of the schools which you attended, including your high school, and you can also look for alumni of any other school. You can narrow down the dates people attended, and then you start getting lists of alumni. I did a quick search for alumni of the UA who attended when I did – between 1992 and 1997 – and it returned thousands of results. Then I began poring through the additional search parameters, which include geographic locations, employers, job functions/departments, major fields of study, and even skills. So let's say I was interested in pursuing a geology career in the Bay Area. I could start my LinkedIn search by looking at people from my university who graduated with a degree in geology and who work in the Bay Area. The site will then provide me with results that meet these parameters, and I would learn about the plethora of career opportunities of my liking, as well as plenty of people with whom I could network in order to begin forging strategic alliances.

- You can designate companies, influencers, and career tracks or industries to follow: This is similar to Facebook (see below), where when you "like" a page, Facebook pushes content related to that page onto your newsfeed. On LinkedIn, when you choose to follow companies, influencers (LinkedIn members who write articles that are posted on the site), or industries, you will find information pertaining to these subjects aggregated in your updates. This is a very effective tool to stay on top of trends in your sector and to identify trendmakers for networking purposes.

- Added benefits with upgraded accounts: As I mentioned, you can join LinkedIn for free, but there are LinkedIn plans for which you pay a premium and in return get certain benefits. Probably the most important perk is "Who's Viewed Your Profile," as I discussed above. Not only does an upgraded account "unlock" the total number of people who have looked at your profile, but it allows you to view their complete profile, as opposed to a modified version of it. An upgraded account has other benefits too:
  - You can view the complete profiles of people who are outside your network.
  - You can send "inmail" to these people (this is different from Invitations to Connect or replying privately to posts in groups).
  - It helps in search optimization: As a premium member, if someone does a search on LinkedIn for a skill, institute, degree, industry, and so on, your profile is automatically moved to the top of the list. So if a hiring manager was scouring LinkedIn trying to find someone with the right skills to invite them to apply for a job, your profile would be among the first they would see in the search results.

Facebook

Although LinkedIn will be your primary social media channel for networking, Facebook also provides value in finding people with whom to network, broadcasting your brand to prospective partners, and gaining inspiration from new sources. At this point in time, most people use Facebook as a way to stay in touch with their friends and family and to share information about their lives. They share jokes and funny pictures (of cats of course), and comment on the state of their existence. Sometimes the posts are friendly, and others are fraught with language and ideas that are inappropriate in a professional context. If you utilize Facebook in this way, just keep in mind that the sixth Pillar of Social Media Networking Principals absolutely applies: No matter your privacy settings, once you post something, there is always the potentiality of someone reading it and possibly taking offense. So be very, very careful what you post, even if you think it is only being seen by your Facebook friends.

I use Facebook in a number of ways to deliver networking ROI:

- I "like" pages that are associated with associations, conferences, people, locations, journals, institutes, and even products: In this way, Facebook acts as an aggregator of news I can use. I gain information about their brands which arms me with inspiration to solve my problems in novel ways or to learn who the trendmakers are in my community. Essentially, I harness Facebook's newsfeed as a way to assess and access hidden opportunities and as way to qualify who the leading professionals are in my field. I can then utilize this information to initiate contact with these individuals, or to make strategic decisions regarding their brands (such as whether I should attend a particular conference or publish in a certain magazine).

- I have two Facebook pages: One, which is under my name, I use for my friends and family to share nerdy humor and to "like" other pages of value. The other is under my company's name "Quantum Success Solutions." People can "like" this page, which gives me the opportunity to promote my brand to potential allies and to post information that would be relevant to those who are interested in my professional services. In particular, my fans can get information from me in their own news feeds concerning career information for scientists and engineers.

  With more and more people using Facebook for both play and professional purposes, you too could create two pages that are separate from each other. One would be for your own personal use (note I don't use the word private here) to chat with their friends and to use as a platform to "like" pages that are relevant to your career goals and hobbies. But you can also have a more public, professional page, which people can "like" and you can use that to send updates. I noticed that individuals were doing this more and more in the political arena. Politicians in the US today often have at least two Facebook accounts: One that usually designates them as a "politician" and is used to raise money and online support, and the other, which designates them with their elected office title (for example, "Member, US House of Representatives"), which they also use to push out announcements about their activities associated with their job.

  So potentially you could have two accounts on Facebook too – one called "Alaina G Levine" and one called "Alaina G. Levine, Scientist." This will be especially useful if you are ever asked by a colleague, boss, or collaborator to be "friends" on Facebook. If you have this other site, dedicated to you as a professional scientist, then you could always say to the boss who wants to "friend" you – "Thank you very much for the invitation to be friends. The account you connected to is the one I use primarily for my family, but I would certainly be happy to connect with you on this other site, or via LinkedIn."

- I post relevant updates: The type and subject matter of the update depends on which page I am posting in, of course, but on my professional Facebook page, it always reflects some aspect of my brand, attitude, and reputation and seeks to deliver value and a ROI for the reader.

- I send private messages: I may notice someone on Facebook, perhaps through a post they made or via a link that returns to their Facebook page and realize that there is synergy in our careers, interests, and goals. I may not have their email address, but I can send them a private message through Facebook.

A final note about Facebook: My sense is that while Facebook right now is used more as a way to socialize with like-minded friends and family, it will become more professionally-focused in the future. So keep up your Facebook activity so that you can stay ahead of the curve as this platform becomes a more integral aspect of your social media networking plans.

Twitter

By now, you probably have heard of Twitter, and perhaps you have signed up for an account and follow a few people or organizations. Twitter is an excellent social media platform for networking and I believe it is both underutilized and often misused for professional purposes. Like other social media sites, it is a tool that you can use to promote yourself appropriately and effectively, establish your brand, and crystallize your reputation in the minds of various publics to gain access to the Hidden Platter of Opportunities.

So as you contemplate incorporating Twitter into your overall networking plan consider the following: You want to ensure that you are adhering to the Seven Principal Pillars of Social Media Networking mentioned above and that you are delivering valuable content that articulates your own talents, expertise, and experience. So please don't tweet about cats and sandwiches and the wind, unless there is a specific purpose that these posts can offer followers.

> **TIP:** Twitter offers a fabulous opportunity to engage in win-win networking. When you tweet about someone's achievement or retweet another person's tweet, you help them achieve their own self-promotion and networking goals.

Instead tweet information that contributes to your community. If you are a postdoc in ecology and want to establish yourself as a thought leader in this field, then it would be appropriate and prudent to tweet articles, images, ideas, and videos that relate to the field of ecology and to your subfields of interest. Tweeting that your cat just died or that the sunset is pretty is just not a good use of your time or the channel.

Some thoughts about Twitter:

- Establish your profile: Just as on LinkedIn, your Twitter profile should include a professional photograph, your name and a title that communicates your brand. You can also include a URL, so if you don't have a personal website, you can include a link to your LinkedIn Profile or your blog.

- Figure out who to follow: Start with your mentors and colleagues and see who they follow. You should follow organizations, conferences, people, publications, products, and companies who you admire and who you think can deliver you value.
- Start tweeting: Go for quality not quantity. You don't have to send 17 tweets a day and they shouldn't focus on your delicious cupcake. Think in terms of the Pillars mentioned above. Your tweets, just like your posts in LinkedIn, should be of value to the professional community and should strategically clarify to others that you are not only a contributing member of that public but are a thought leader as well. So what should you tweet?
  - Articles that are related to your profession.
  - Videos or pictures that are related to your profession.
  - General ideas, thoughts, quotes.
  - Questions – this is especially useful at a conference or within an organization.
  - Expressions of gratitude, plaudits or congratulations relating to specific people or events: I recently attended a teleseminar which I found incredibly helpful. Afterwards, I sent a tweet thanking the presenter and complimenting her on her valuable presentation. I incorporated her twitter handle in the tweet so I knew she and her followers would receive it. She responded by favoriting my tweet, which consequently exposed me to her network.
- Tweet strategically: Use @ signs in your tweets to get the attention of people and organizations who would be interested in your specific tweets and more broadly, your work. And use hashtags too (#), especially at conferences or as it relates to a conference.
- Retweet away! If someone tweets something you find interesting and that you think will be of value to your followers or to the greater community, then feel free and retweet. You can and should also favorite tweets. By doing so you may just gain a new follower yourself, because you are bringing yourself to the attention of the person whose tweet you retweeted. But more importantly you are delivering on the promise of networking – that you are not just taking what you can get from the person but instead you are contributing something to their wellbeing as well. In this case, you are promoting them and their work to your own network, which will raise their visibility in the community. They will appreciate this greatly. And you can utilize this as a springboard to send them a private message (PM) about setting up an appointment to speak about mutually-beneficial projects.

But before you retweet... read the original tweet and its link! The last thing you want is to retweet what appears to be a link to an interesting story about frog biodiversity and instead end up tweeting a link to a

> **TIP:** Among its many benefits associated with strategic networking, Twitter offers you a chance to help other parties expand their own networks and amplify their brands and reputations by tweeting about them and retweeting tweets that originate from them. People appreciate this gesture!

pornographic website. So for every tweet that you get that you want to forward to your followers or make known to others (by using @ for twitter handles or hashtags for people following a certain thing), clarify what the link actually links to. Your reputation is at stake.

Probably the most valuable way that you should engage and use Twitter is for gaining information about conferences. As I stated in Chapter 6, Twitter use at conferences is becoming more and more the norm. As a result, more people are engaging Twitter to discover and publicize hidden opportunities and even create their own opportunities at meetings. You can take full advantage of this. Some thoughts about using Twitter at conferences:

- Check the Twitter feed before you attend. People often start tweeting about the conference at least a week or more in advance. Furthermore, if you are following the conference hashtag, you may find out information that is not found anywhere else or can assist you in making myriad conference-related decisions that range from what sessions to attend, to accessing evening mixers, to finding people with whom to have coffee while in attendance.
- Tweet about sessions. If you are participating in a terrific session or listening to a great presentation, tweet it and include the conference hashtag and the speaker's Twitter handle.
- Tweet about logistical info that others will find helpful. For example, at a recent conference in Minneapolis, I noticed a few tweets about using light rail to get from the airport to/from the hotel area in the city. As a result of those tweets I found easy and cheap transportation to the airport on the way back.
- Tweet about networking mixers and related events. Lately I noticed that Twitter is being used as the sole channel for advertising certain evening events that are unaffiliated with the sponsoring association. At a recent Ecological Society of America Conference, there were numerous mixers organized by alumni and special interest groups that were promoted only on Twitter via the conference hashtag of the conference. If I wasn't following the feed, I wouldn't have known about them otherwise and would have missed these strategic networking opportunities.
- Be courteous. I have noticed that many organizations have established social media guidelines for attendees and have even printed them in their conference programs. One of the major points that is usually

conveyed is the need to be respectful of someone's privacy at a meeting. This means don't tweet something you just overheard two people discussing in a private conversation. Only tweet information that is public – knowledge you gain from someone's public presentation or poster.

- Remember that anything you tweet is completely accessible by the public. So be careful before you tweet something, especially if you don't have your facts straight. At one conference one person launched a tweet dialogue about how expensive the Wi-Fi was at the convention center. But what this person didn't realize was that the Wi-Fi network she was referencing was not the Wi-Fi that the association had negotiated to provide for free to conference participants. But once she tweeted the incorrect information, it started a tidal wave of nasty comments to and about the organization, which in the end were completely unfounded. The organization couldn't rally those negative perceptions in – the misinformation was already out there, and it was due to one person who tweeted without thinking or checking their facts first.
- Don't tweet about negative things. Even if I was in a session that was boring and I considered a waste of my precious time at the conference, I would never tweet "This sessions sucks! #conference2014," and you shouldn't either. Remember you are on a public stage when you tweet and we only want people to associate you with positive ideas. Tweets that border on negativity, bring others down, or communicate faulty or even potential libelous untruths will only serve to damage your reputation and conceal potential collaborative opportunities. So related to this issue, keep your tweets clean. Don't curse, and don't make potentially offensive remarks.
- Send Private Messages. If you see someone tweeting about something at the conference and you want to meet them, you can always send them a Private Message through twitter.

Other Social Media Sites

There are plenty of other online channels that you can utilize for networking purposes. The following is only a small representation of all that is currently available. Keep in mind that new platforms are added constantly and part of your networking strategy is to stay abreast of what is being launched and how it is being used by members of your community.

- Google+: This site involves "Circles" of networks. I predict that more professionals will utilize this site in the near future and therefore it will become more valuable for networking. Join it now and stay tuned.
- Instagram: This platform is mostly used to share pictures, and is often integrated with Facebook and Twitter. I have noticed more people

posting pictures from conferences on Instagram and then tweeting the Instagram picture.

- Pinterest: This site allows you to share websites, articles, videos, and pictures that you find interesting, as if you were "pinning" them to a bulletin board. I predict this site will also become more useful and utilized by STEM professionals in the near future.
- ResearchGate: Specifically for researchers in academia, laboratories, and government agencies, this site allows you to find and connect with people who are pursuing similar research objectives.
- Frontiers: Another site designed for the research set, Frontiersin.org promotes itself as a "community-oriented open-access academic publisher and research network." You can join and connect with researchers in your specific field and also read and contribute to discussion on articles and journals.

> **TIP:** You can often link your profiles on different social media sites, for ease of use and to save time. For example, you can post an update on your LinkedIn profile and simultaneously tweet it, and you can tweet something and have it also appear on the Facebook newsfeeds of those who "like" your page.

- Social media sites run by professional societies, or those related to publications: Check out if your membership in your association includes the ability to access an exclusive social media site. This could be a very useful channel. And on the websites of publications like ScienceCareers and NatureJobs you can also find solid social media networking opportunities.

Launching a Blog

Many early-career scientists and engineers often wonder whether they should have a blog. A blog can offer great opportunities for self-promotion: it can help you build a following in your community, it can establish you as a thought leader in your field, it can help you sharpen your writing skills, and it can even help you land a job.

But writing a blog can also be a time-consuming enterprise, so before you embark on doing so, there are a number of decisions you need to make:

- What are your objectives for launching one? Is it to establish yourself as a leader in your field, or to share information about others' contributions to the discipline? Or do you aim to expound on your love of bacon or whittling wood jewelry?

- What will you write about?
- Where will the material come from?
- What form will the material take – will it include text and pictures, and videos too? Or perhaps powerpoint presentations?
- How much time each week can you devote to the blog?
- How often will you post updates? If you don't adhere to a certain level of frequency in your posts, you may find your readership dissipate and slip away, defeating a major reason to have the blog in the first place.
- What platform will I launch it on? Most people use wordpress these days, as it is a simple and free to use.

Some thoughts about blogs, especially if you are using this as a platform to promote yourself in your area of expertise:

- Don't repost anything without attribution.
- Don't repost pictures without permission and/or attribution, depending on the source.
- Use a blog title and URL that is catchy and easy to remember.
- Use "tags" so people can more easily find your blog.
- Promote your blog, via LinkedIn, Facebook, and Twitter – you don't have to make a LinkedIn announcement for every entry, but you can post occasional notes about interesting blog entries. Additionally, if there is something special occurring in the news and your blog post is related, then this would be considered potentially valuable information for others to have. Share it via your other social media channels.
- Embed a form in your blog that allows you to collect email addresses of people who visit and voluntarily supply them.
- Connect your Twitter feed, LinkedIn profile, and Facebook account to your blog, so when others read it they can easily share it on these channels themselves.
- If your blog is directly related to your area of expertise, put the blog URL on your résumé, LinkedIn profile, Twitter account and possibly even your business card.

Once you get used to blogging and even build a following, you can leverage this expertise as a way to write for established blogs on websites and for publications that relate to your industry. For example, you can offer to guest blog on science websites, for STEM organizations, and even magazines, like *Scientific American*. *Scientific American* in fact welcomes blog posts, as do many science magazines, and this is an excellent opportunity to reach a new audience and expand your network.

Your Klout Score

Klout.com is a website that tracks your activity in social media spaces to provide a measurement of how influential you are in the community. You sign up for an account and link all of your social media sites to your Klout profile. Klout then keeps tabs of your posts, and more importantly, others' responses to your posts. So the more you post and the more your posts elicit engagement from others, the higher your Klout score becomes. Your online interactions are ranked from 1 to 100, with 100 being the highest. A higher score means that you are a bigger player in cyberspace, and more importantly could also lead others to perceive that you are a thought leader in this arena.

So at this point I know you are probably thinking to yourself: Why should I care what one website right now says is my "measurement" of online influence? According to a *Wired Magazine* report, Klout is just one tool that potential employers and partners use to quantify your influence. And since networking involves convincing others of the value you can provide, your level of influence in a community speaks volumes about that value. Some industries take the Klout score so seriously that if your score is low, not only will you not be able to secure a job in that sector, but you won't even make it to the interview. Now admittedly, these jobs mostly exist in the marketing, public relations, advertising, and of course social media arenas. But take note: Klout is a harbinger of things to come. I suspect in the near future all of our social media activities will be measured in some way, either by Klout or some other channel, and it will be vital in any industry to ensure that your score is high because it acts as a representation of how valuable you are and how much value you contribute to the community.

We can learn a lot about social media networking from successful politicians such as President Barack Obama, who many have stated won the 2008 election as a direct result of his social media strategies and tactics. Take a page from him and others: Use social media the right way and it can help you win a campaign (or a contact, or a job!).

Chapter Takeaways

- Having a strong social media presence, which involves active participation on multiple social media sites, is an absolute necessity for career advancement and must be part of your overall networking strategy.
- Follow the Seven Principal Pillars of Social Media Networking.
- Information should go both ways.
- Be professional in content, communications, attitude.
- Anything that is online will be there forever – "Cyberspace is Foreverspace."

- More employers do simple Google searches and more complicated social network searches than ever before.
- Know what is online about you and if it is something negative, try to either remove it or clean it up, or be ready to address it openly in job interviews or other networking circumstances.
- Link your various accounts on social media sites for ease of use and to save time.
- Pay it forward – just as you want to utilize social media for networking and reputation amplification, so do your colleagues. So promote their work via tweets and updates – they will appreciate it and it goes a long way to building those lasting win-win relationships.

Notes

i.   From an email request by Yumi Wilson, LinkedIn for Journalists, May 25, 2014.
ii. From a teleseminar by Yumi Wilson, LinkedIn for Journalists, August 20, 2013.
iii. Ibid.

9   The Networking Continuum

Now that you have ideas and (hopefully) inspiration about how you can grow and cultivate your networks for your own professional advancement and that of your organization and your networked contacts, let's see what we can do to keep the momentum going throughout your career(s).

This last chapter is meant to serve as a way to help you put all of these ideas, tactics, and strategies relating to networking into practice. Our goal is to maintain our networking, keep our networking actions fresh and dynamic and constant, in order to nurture your networks and expand into new ones. And we also want to ensure that our networking activities become less of a chore and more of a natural and fun action that you don't even realize is "networking" when you are doing it. This is where you get some homework!

Homework assignment 1: Start networking! Whether you are in your postdoc appointment, grad school, college, high school, or yes even middle school, start speaking with those around you, your friends, your bosses, your teachers, your mentors, your parents even, and ask questions about their careers. Learn as much information as you can about the different types of fields that spark your interest and look for ways NOW to access Networking Nodes that can help you and your networking connections now and in the future. And remember: It's never too early to network.

> **TIP:** It's never too early to network.

As I have noted in this book, networking must have a place in all phases of your career. And I hope it is also clear that it is never too early to start high-impact networking. You simply can't afford to wait until you graduate to start networking to find a job, because by then many hidden opportunities might have already been offered to your competition. And since our guiding strategy is to build long-term value exchanges with every contact in our network, the sooner you begin to engage people about their work and learn how you might be able to aid them, the sooner

Networking for Nerds: Find, Access and Land Hidden Game-Changing Career Opportunities Everywhere, First Edition. Alaina G. Levine.
© 2015 John Wiley & Sons, Inc. Published 2015 by John Wiley & Sons, Inc.

you will learn how to advance your career into new directions and land your dream job(s).

Here's an example of someone who understands this principle: In 2013 I was invited to speak at Caltech by their Postdoctoral Association. The promotion of the event landed in the email inbox of Serina Diniega, an applied mathematician who works for the Jet Propulsion Laboratory (JPL), which is very closely linked with Caltech. By coincidence I had recently written a story for *Science Magazine* about how postdocs can better market themselves and had featured this young woman in the article. Of course, we had connected through networking: She read about my interest in finding sources for this article on a Facebook group to which she belongs, and reached out to me. And as it turned out, we had both attended the UA (although at different times) and as such we realized our networks overlapped in various ways. When she saw that I was going to be on campus, she reached out to me to welcome me and offered to give me a tour of JPL. I was so excited by this prospect. So the morning I was to fly back home, Serina picked me up at my hotel, drove me to JPL where we had breakfast, and then gave me a fabulous tour which included a view of the control room where all the Mars Landing Mania had occurred.

I truly appreciated that she took time out of her busy schedule to give me such a personal tour, and of course I sent her a thank you note. But one aspect of the whole experience particularly impressed me: Serina had invited her protégé, a community college student, to join us for breakfast and the tour, knowing that the protégé was interested in science writing and I am a science writer. The protégé has been spending time at JPL doing her own research project. During our meal, and later as we traversed the campus, the student asked me various questions about my career, which I was all the more happy to answer. I gave the protégé my business card and told her, as I did Serina, to stay in touch and keep me informed of her progress. And then a funny thing happened. The community college student emailed me a thank you note to express her appreciation for taking time out of MY schedule to speak with her. This whole exchange demonstrated some very important issues:

- Serina's gracious invitation to me helped solidify our partnership even more than before. She not only offered me something of value (the tour and time spent getting to know more about her), but she also amplified her brand and attitude in my mind and, thinking forward, will also ensure her reputation is carried in a positive way to decision-makers. Because of our meaningful engagement at the lab, I am more likely to go to her with ideas for collaborating or when I hear of jobs and other opportunities which may be of interest for her.
- Serina's own initiative to help her protégé expand her career horizons by inviting her to meet with me and ultimately network with me.

In doing so, she subscribed to the principle of paying it forward in networking, which can be extremely powerful for all involved.

- The community college student's boldness to actually go out and get an internship at one of the premier research laboratories in the world. Most college students, while enterprising, may not have the guts to ask for and apply for such a prestigious opportunity.
- The community college student's creativity in leveraging her own network to get the position at JPL: She probably can't even legally drink yet, but she understands the importance of networks and networking to access hidden and amazing career-changing opportunities.
- The student's understanding of the importance of networking, especially early in one's career: In her email to me, she expressed interest in staying in touch with me and helping me with my projects. I am extremely impressed that this student not only had the wherewithal, guts, and fortitude to send the email, but to phrase it in such a way that she envisions we can help each other down the road. This girl is a networking ninja!

> TIP: It's never too late to network.

This whole exchange is a terrific example of this very important point: It is never too early to start high-impact networking! So get to it! Do it now!

Homework assignment 2: Reach out to colleagues with whom you haven't spoken in a long time.

No matter how you met them, whether it was in grad school or at a job or on an airplane, even if it's been quite a while since you last spoke with them, (or perhaps you never spoke with them after the initial meeting!), it is never too late to re-initiate contact and start building a relationship with them.

As I mentioned in the Introduction, in 2001 I attended the annual meeting of the American Physical Society and hung out in the press room, hoping to meet journalists and other communications professionals. I had just come from a stint handling PR for the UA Physics Department so my interests were both timely and genuine: I wished to expand my networks and improve my networking skills. One day at lunch a bunch of us were sitting around the table chatting when it was revealed that one of the party was a physics professor at a prominent university in Europe, in a country that I had always desired to go to. We spoke casually and I may or may have not followed up. In fact, I didn't follow up again until 2006, when I sent him an email about the potential of coming to his university to speak and to perhaps work with his students. I might have been hesitant to email him since what appeared to be a long time had passed, but I said to myself, "what the heck? – it can't hurt to reach out after these years." The worst that could happen would be he would say no or not answer my email at all. To my delight and surprise, not only did he respond to my email but he was happy to assist me in my quest and provided me with ideas and even references of people to whom I could turn for my project.

I have found this to be the case with many people with whom I may have lost contact, or didn't follow up at the time of the original engagement. I may have their business card on my desk for years and don't reach out, but the beauty of networking is that once the contact is made, the door (for the most part) is always open for you to return and see how you can assist each other with your projects and goals. So don't hesitate and don't let those business cards go to waste – reach out to them now!

Homework assignment 3: It's never too late to reach out to people you have wanted to contact but just haven't.

Perhaps you saw Dr. X speak at a conference five years ago, or read Dr. Y's paper in *Science* six months ago. Or maybe you read about Dr. Z's achievements in *The New York Times* a decade ago. All of these people were appropriately and successfully engaging in self-promotion activities as I discussed in Chapter 4, so their aim was (and continues to be) to expand their own networks through these promotional channels. Even if they wrote the article 10 years ago and you found it fascinating, either when it was first released or more recently when you came across it, you can still reach out to them to make an appointment for an informal conversation to learn more about their work and explore how you could contribute. They will be tickled that even after all these years their message was appreciated and that others found their accomplishments impactful. While I was writing this book, I reached out to someone who is mentioned here to ask his permission about using his quote and message in the book. I saw him speak years ago at a conference, and when I told him that, he immediately expressed appreciation and gave me permission to quote him. He was probably touched that his message was so impactful to me that I remembered it after all this time. And in his message back to me, he offered to help me with my projects and looked forward to chatting further.

Homework assignment 4: Recognize the richness of your current networks and your access to certain Networking Nodes as providing you lifelong insight and assistance.

The people around you now, your fellow colleagues, students, bosses, mentors, protégés, are all part of your permanent network. Just as you advance in your career, they too are advancing in theirs. You will have needs for inspiration and leads for career opportunities and they will too, and given that you already have a connection with them, you will generally always be able to reach out to them to suggest ways in which you can help each other, now and in the future. You will always have these ties that bind you together. So when you think about it, you have a pretty healthy network from which to grow even more networks and strategic win-win partnerships. And this is what I mean when I say the networking relationship ends only when one or both of you drop dead. Because even 22 years from now, you can still reach out to Dr. X and ask if you can meet for coffee or have a Skype appointment to update her on your work and learn what she's doing so you can possibly assist her in some way. I still stay in contact with all of my previous bosses, even the one I had as a student worker. I have known her for 18 years and we still occasionally put our heads together to see how we can help each other. And thanks to social media, specifically Facebook and LinkedIn, it is even easier for me to reach out to old high school chums and renew friendships and craft new partnerships. So remember, you are networking *now* for the future. Keep those connections current, and when someone physically leaves your presence, perhaps to join a new lab or take a fellowship in another country, that doesn't mean the relationship has ended or the Node has crumbled. Stay in touch! An easy way to incorporate this in to your networking plan is to make it a priority to reconnect with one of your former colleagues or peers at least once a month.

Homework assignment 5: Realize that Networking Nodes can be anything and can happen anywhere, at any time. In fact in many regards you can make any thing, person, event, and even object into a Networking Node.

The most obvious example of this principle is this very book. The information I provide here all stems from my own networking, both from tips I have learned from others or have realized myself, along the way throughout my careers as I endeavored to expand my perspective and horizons. In writing the book, I consulted with colleagues from my present, not-to-distant past, and ancient times (about 25 years ago). In a very real sense, this book served as a Networking Node for me, because it enabled me to start and continue conversations with colleagues, new and old alike. My expectation is that it will continue to be a Networking Node for me, both personally and professionally, as I hope to leverage its existence as a way to start conversations with new people, learn about novel ways to solve my problems, gain access to Hidden Platters of Opportunities that I couldn't even imagine existed, and of course, to provide value to other nerds (and non-nerds too) who want to sharpen their own networking abilities.

But for you, the reader, this book also can serve as a Networking Node. You can also use it as a way to start conversations with those with whom you wish to connect, talk with like-minded nerds about strategies for networking in different sectors and ecosystems, and ultimately approach your day-to-day problems from new perspectives and with new motivation.

So look for avenues for you to turn a seemingly ordinary conference, article, reception, coffee appointment or email into a Networking Node, through which multiple people can gain high-impact networking ROI. Here is a fabulous example of a potential Networking Nodes you could either create or leverage: If you are giving a poster presentation at a conference, your poster and your presence in front of your poster serve as a Networking Node. People are walking the poster farm, looking at different scholars' works, and when they come across you and you smile and welcome them into your poster space with a handshake, you have now created a perfect opportunity for you and the other party to network. Take advantage of this golden opportunity!

Homework assignment 6: Get on LinkedIn immediately: Launch your professional online presence in this professional marketplace.

Create your profile, join groups, start asking questions and contributing to conversations and start connecting with people who are interesting. Create those digital trails to amplify your brand, attitude, and reputation. And here are two tips to really engender a profitable networking continuum: Once a month, look through the membership roster of one of your groups and take note of people who you may want to contact. Similarly, once a month, take a look at one of your LinkedIn connections own connections and make a list of people with whom you think you could build mutually-beneficial alliances. Start emailing these people. These two steps will open up whole new worlds of contacts and Networking Nodes!

Homework assignment 7: Look for opportunities to weave your career goals and your networking goals together.

Whether it is by networking with individual professionals in the sector in which you desire to work, or joining groups on LinkedIn, or going to strategic conferences, keep your eyes open to the possibilities of how your networking activities can contribute to your career objectives, and how your career outputs and aspirations can enhance your networks and networking ability. A good point to remember is that when I hire you for a job, I not only base my decision on the sum total of your skills, experience, and expertise, I also am thinking holistically about the decision – you have extensive networks which can provide value to my organization, through you, and your level of expertise is even greater because you have access to so many other diverse experts. So in many ways, my choice to hire and invest in you as an employee is made because of your abilities AND the abilities of those you have access to who can help you when you are stuck on a problem. It is easy to see, with this type of example, how your networking goals are tied to your career goals – the more you network the better you can become in your job. And furthermore, thinking more broadly, the more you network, the more you can become better at career planning for the future. If you can start to draw parallels between your networking objectives and achievements and those in your career as well, you will be able to more smoothly move forward in your profession in the direction you desire.

Homework assignment 8: Think entrepreneurially, always, in all ways.

I have mentioned this multiple times in this book, but it bears repeating: It is absolutely critical that you always think entrepreneurially about your career and, by default, your networking activities. You have to approach your professional progression and that of your colleagues in your networks with an entrepreneurial mindset – how can I make this project better? How can I inject value in a new way into your organization or team? How can I solve the problem differently to move the field and organization forward, faster? By keeping this mantra in the front of your mind with every action you take, you will find that your networking will deliver greater ROI for everyone, because as you engage someone and learn about their craft you can become better at their craft. Remember my uncle – the entrepreneur, genius, and garbage man-questioner – who always looked to people with diverse backgrounds to seek answers for his own problems in his field and company. This attitude of welcoming insight from diverse publics must be part of your networking strategy and it should be part of your career advancement strategy as well. You can practice being an entrepreneur as you network by looking for new Networking Nodes you can create and launch, such as a new group on LinkedIn or an informal gathering at a conference. You can also train your brain by seeking out and designing your own networking opportunities closer to home. For example, if your department doesn't have a journal club or a colloquium series, consider starting one. Or suggest an internal awards program for your company and offer to organize it. There are many, many ways you can launch your own networking enterprise, channel, and opportunity, and as I have written, by simply doing so, you are establishing that you are a leader in this arena, elucidating your brand and attitude, magnifying and propelling your reputation to new contacts, and strategically creating a way for you to both reach out and network with new people, and for new people to reach out and connect with you. Just like networking itself, everyone wins when you think and act entrepreneurially in your networking and career development activities.

Homework assignment 9: Enjoy your networking!

This is fun! This is a pleasurable experience! As a networking master, you get to speak with people about your passions and learn about their own desires and goals. What could be better for a nerd? So go forth and ...

- Practice your networking: Don't be shy; you CAN engage in high-impact networking. You can overcome your shyness or introvertedness to reach out to potential allies. The more you do it the easier it will become and I guarantee it will become second nature!
- Be proud of your networking: The fact that you understand, appreciate, and will now leverage your networks to move your career, field, scholarship, and those of your contacts, forward is commendable.
- Enjoy your networking: This privilege of exchanging ideas and inspiration is something that very few people take advantage of and experience fully.
- Profit from your networking: There is nothing wrong with bettering yourself and those around you, and networking holds the key to doing so.

- Pay your networking success forward: The more you help those in your networks to attain their own professional victories, via that win-win partnership we have been discussing throughout the book, the more you will discover you can advance in new avenues too.

And finally, celebrate your nerdom! Improving your skills in networking does not mean your quintessential character and core, which may be defined by your nerdiness, has to disappear. Rather, you can still be a nerd and be a master networker! I wouldn't want it any other way, and I am sure you would agree.

I will end this book by saying that I am very excited about what your future holds for you. You've taken an excellent first step in exploring new careers and meeting amazing people with whom you can partner. I won't wish you good luck, because you don't need it and you certainly don't need me to suggest you need it. You have all the power, intelligence, and abilities you need to move your career forward and into any industry, sector or universe you please. Follow your interests, improve your skills, help those around you, and keep networking. And relish the ride – because you are in for some amazing experiences!

Index

Networking for Nerds: Find, Access and Land Hidden Game-Changing Career Opportunities Everywhere, First Edition. Alaina G. Levine.
© 2015 John Wiley & Sons, Inc. Published 2015 by John Wiley & Sons, Inc.